WAR
STORIES

Also by Peter Snow

Leila's Hijack War

Hussein: A Biography

Battlefield Britain (with Dan Snow)

Twentieth Century Battlefields (with Dan Snow)

To War with Wellington

When Britain Burned the White House

The Battle of Waterloo Experience (with Dan Snow)

Treasures of British History (with Dan Snow)

WAR STORIES

From the Charge of the Light Brigade to the Battle of the Bulge and Beyond

Peter Snow &
Ann MacMillan

PEGASUS BOOKS
NEW YORK LONDON

War Stories

Pegasus Books, Ltd.
148 West 37th Street, 13th Floor
New York, NY 10018

First Pegasus Books hardcover edition September 2018

ISBN: 978-1-68177-852-5

10 9 8 7 6 5 4 3 2 1

Printed in the United States of America
Distributed by W. W. Norton & Company, Inc.

To the men and women in this book and the countless others whose stories have yet to be told

Contents

INNOVATION

MEDICINE

WAR AT SEA

WAR IN THE AIR

ESCAPES

COMPASSION

Introduction

We hope this book will give the reader an intimate glimpse of men and women who have risen to the challenge of war with acts of great heroism and humanity. Most of them are ordinary people swept up – often by chance – in the most extreme ordeal humans have to endure: their actions are the stuff of legend. Our careers as journalists have given us a powerful insight into the great engines that drive history. The leadership summits, the treaty signings, the handshakes between titans: we've been there reporting on many such occasions. But just as compelling are the stories that have changed history on the ground, men and women who have pushed the barriers of bravery, suffering and sheer terror beyond the imaginable.

There are hundreds of people whose experiences in wartime have been so remarkable that they've been recounted again and again. There are thousands more whose stories have been pushed aside in the passage of time. In this book we choose from the best of both. We've always been captivated by the well-documented exploits of rascals like America's brazen turncoat Benedict Arnold who switched sides and joined Britain in 1780, and the inspiring humanity of Sir Nicholas Winton, famous for rescuing children from the Nazis. But most of our tales are less well known: like the torment of Lady Florentia Sale, held hostage by Afghan rebels in 1842; John Rabe's struggle to save Chinese lives in the brutal Japanese assault on Nanking in 1937; and Dr Sumner Jackson who treated fugitive Allied airmen at the American Hospital in occupied Paris during the Second World War. Some of our accounts will surprise readers because they are little known or in some cases scarcely mentioned till today.

The twenty-eight individuals and three couples we have selected span nearly three centuries from the terrifying odyssey of John

Bulkeley of HMS *Wager* along the remote coast of Chile in 1741 to the heart-stopping escape from Syria of Ahmad Terkawi in 2015. Their stories take us to battlefields all over the world – from China, Burma and Afghanistan through the Middle East to North Africa, Europe and North America.

All stories are based on primary source material. Most of the people we feature have written their own accounts of their experiences or been written about by others in diaries, letters or autobiographies. We have first-hand accounts too. We have spoken to a number of people still alive at the time of writing – people like Corran Purdon who survived the St Nazaire raid in 1942 and Joe Schlesinger who told us how he was rescued from the clutches of the Nazis in 1939.

This is not a list of unconnected tales. All are interrelated. All are haunted by the spectre of war. We have weaved them together by themes because the individuals we've chosen seem to us to highlight one particular human quality that characterises the nature of warfare. For example, the opening part on Courage includes the apparently fearless academic from Maine, Joshua Chamberlain, who insisted on leading his men from the front in battle after battle in the American Civil War. Others like Augusta Chiwy, whose cool-headed nursing in the midst of the Battle of the Bulge in 1944, come under the heading of Medicine. The ingenious trickery of the Second World War German pilot Franz von Werra makes his one of war's most audacious Escapes. Many of our protagonists could comfortably have come under several of our headings, but we have chosen the one that we feel most suits each person.

Researching and writing about these exceptional people has brought us very close to them. We find ourselves admiring even the scoundrels. We hope they all come as alive for the reader as they do for us.

COURAGE

LIKE MOST PEOPLE in today's world we have never actually fought in a war. We have both seen people fighting in wars and we have often wondered how we would behave in battle – whether we would turn and run or have the courage to stand and fight. History is full of examples of people whose initial fear has given way to courage – courage that fortifies men and women to do astonishing things way beyond the demand of simple duty – from men like Leonidas the Spartan king who with his tiny force held off thousands of Persians at Thermopylae in 480 BC to women like Edith Cavell, the British nurse in Belgium in the First World War, who paid with her life for smuggling scores of fugitives to safety. Occasionally courage can verge on recklessness. British paratrooper Colonel 'H' Jones may have shown unwise leadership by charging off ahead of his men at Goose Green in the Falklands in 1982 but his courage was not in question.

Although today's wars are fought at much longer range and the opportunities for people to show courage fighting for their lives at close quarters are now much rarer than they used to be, there is no shortage of examples. We've singled out three individuals who seem to us to personify real valour. Each of them had to confront extraordinary challenges and each measured up. The most recent was a young tank driver in the Iraq War in 2004 whose courage won him the Victoria Cross. A century and half earlier a university professor joined the Union side in the American Civil War and his extraordinary bravery earned him the rank of general in only two years. But we begin back in 1854 with a young cavalry captain in the Crimea who led his men in the Charge of the Light Brigade, as great a test of a soldier's courage as any in history.

I

Edward Seager

One of the 600, Balaclava, 1854

EDWARD SEAGER WAS the image of a dashing hussar. Tall, dark, well-built with a generous sloping moustache like an upturned V, he wore a glittering braided uniform that outshone any other regiments in the army. On 25 October 1854 at Balaclava he took part in one of the most famous cavalry charges in history. In that shamefully misconceived and bloody disaster he was one of those gallant men described by Alfred Tennyson in his powerful poem, 'The Charge of the Light Brigade': 'Theirs not to reason why, Theirs but to do and die.' And Seager was lucky: he rode into the 'valley of Death' and survived to tell the dreadful story of what happened.

Just hours before the charge Seager had watched Jerry, his favourite horse, die after he had been cooped up in the hold of the ship bringing the regiment 2,500 miles from England. It was 'piteous to see him', Seager wrote home to his brother: 'The effect of being in so confined a place . . . gave him disease of the lungs.' Jerry 'had been one of the best chargers in the regiment'. Seager now had to rely on his second horse, a six-year-old mare named Malta, which he'd bought just before he sailed. Little did he know it, but the day ahead would see the mare and her rider plunge into a ghastly maelstrom.

Seager checked his equipment. He had the slashing sabre of a light cavalryman and – like most hussar officers – a revolver. Slung on his belt he carried an ornate pouch with some precious contents. He later told his wife, Emily, that through all those dangerous days his bag contained 'the darling children's picture, my dear mother's present [Prayer Book and Testament] and lots of letters'. In his jacket pocket he had 'your letter containing dear little Emily's hair'. Jane Emily was the eldest of their seven children. Around his neck, he told his wife, he had 'the dear locket you gave me in Exeter'.

He had come a long way from his peaceful home in Devon to the shores of the Black Sea on the southern edge of the Russian Empire. The Crimean War was nearing its climax. Britain, France and their Turkish allies, whom Seager, along with many of his fellow soldiers, regarded as little short of useless, were forced into an unlikely alliance against Russia. Britain and France had little in common with the Ottoman Empire and its Muslim rulers but they did not want to see it disappear, thereby leaving a power vacuum that Russia could exploit. In 1853 Tsar Nicholas I had launched his armies south into the Balkans intent on seizing parts of the Turks' ailing Ottoman Empire which stretched from the Middle East up into the Balkans. The Russian attack had failed and Britain and its two allies had by the summer of 1854 pushed the Russians back north and east into the Crimea. They'd won a victory at the Alma River on 20 September and were now encircling the great Russian fortress of Sebastopol. But in mid-October the Russians moved forces up to threaten the allied supply base at Balaclava six miles south-east of Sebastopol. A major battle was inevitable.

Captain Edward Seager was forty-two. He'd wanted to join the army ever since as a child he'd watched the 18th Hussars parading in the streets of Liverpool. He ran away from home at the age of seventeen and became a soldier, but his furious father, a prosperous brewer, bought him out and elbowed him into the family business. Three years later the undaunted young Edward enlisted again; he changed the spelling of his name and didn't write home for four years. He was a fine horseman and by 1854 he was adjutant in the 8th Hussars, a regiment of light cavalry that had been earmarked for the Crimean War. Seager was soon ordered to embark with his two horses, Jerry and Malta.

On 2 May 1854 Seager boarded the troopship. 'Poor Emily,' he wrote to his brother, 'was on the pier as we left and I saw her standing there for a long time. Take care of her and my beloved children.' The ship ran into a gale in the English Channel. There were eighty-two horses aboard and the rolling of the vessel made them frantic. Each time the ship heeled over, every horse dashed forward trying to get free, making a fearful noise with their hooves and some of them, as Seager put it, actually 'screaming with terror'. Three horses died during the night. 'We threw them overboard the next day.' Seager's mare took it all very quietly but Jerry 'was so frightened that he got his feet in his manger, struggled violently and at last fell down'. His legs were quite badly cut and Seager had to enlist a few men to help him heave Jerry to his feet again. The horse survived for only another five months.

Another, much more serious crisis hit the regiment while they were being held in Bulgaria before joining the fighting. 'A far worse enemy than the Russians is amongst us,' Seager wrote home. Disease – cholera, fever and dysentery – was prompting 'a funeral nearly every evening'. Finally, to Seager's relief, the regiment was shipped across to the Crimea and landed with some difficulty on a beach. On 20 September it helped scatter a Russian force waiting at the River Alma to block their advance on Sebastopol. By the end of the month Seager's 8th Hussars were posted outside Sebastopol guarding the approaches to the alliance's supply base at the port of Balaclava.

They expected to do more than just protect supply lines. Cavalry, both heavy and light, played a key role in nineteenth century warfare. The less heavily equipped cavalry made natural scouts. Both types of mounted troops could be thrown against enemy infantry lines, and, in the moment of victory, chase down those who fled.

The 8th Hussars formed part of a light cavalry brigade of some 600 men and their horses led by Lord Cardigan, brother-in-law of the overall cavalry commander Lord Lucan. The two men were known to be fierce rivals who couldn't stand each other. Seager had little time for either of them. He reckoned Lord Cardigan had made a hash of the beach landing in the Crimea, and Lucan he described as 'Lord Look-on', hopelessly cautious. In the morning of 25 October a large force of Russian infantry and cavalry seized a ridge north of Balaclava. They were only stopped short of invading the base by the prompt resistance of the 93rd Highlanders and several charges by the Heavy Brigade of cavalry. To their chagrin the Light Brigade were not called on to help with this counter-attack on the retreating Russians. The 8th Hussars had to sit in their saddles and watch. But not for long.

When the Russians overran the ridge earlier on, the British retreated, leaving precious guns behind. Lord Raglan, the commander of the whole British army, ordered Lucan's cavalry to advance and stop the Russians carrying off the British weapons. It was the first order in what would become the most infamous chain of commands in military history. Raglan gave his order to Captain Louis Nolan, who passed it on to Lord Lucan, who finally handed it to Lord Cardigan. Raglan's order was for the cavalry 'to advance rapidly to the front – follow the enemy and try to prevent the enemy carrying away the guns'. This was almost certainly intended only to trigger a cavalry advance to recapture the British guns on the ridge. But when Lucan asked Nolan where he should attack, Nolan replied, gesturing at the valley beyond the ridge, 'There, my lord, there are your guns,' and he pointed to a line of Russian guns way down at the far end of the valley. Lucan then took the order to the next link in the chain, Seager's brigade commander, Lord Cardigan. Cardigan was astonished: the valley he was being told to enter was about a mile long with Russian guns and riflemen along both sides of it and a line of Russian guns at the end of it. 'Cannon to right of them, Cannon to left of them': Tennyson's reference to the 'valley of Death' was no exaggeration. Cardigan exclaimed that it was lunacy, but Lucan insisted. The two antagonistic brothers-in-law argued for a time but orders were orders: Lucan was a two-star general passing on to Cardigan, a mere one-star brigadier general, the orders of three-star General Raglan. Whether or not Lucan or Nolan had misinterpreted the orders deliberately or by accident, the orders had to be obeyed.

Seager and the men of the 8th Hussars had spent the morning awaiting the orders of their commanding officer Lieutenant Colonel Frederick Shewell, whose face was almost invisible behind a great thicket of hair that hid all but his eyes. Shewell wasn't popular: he was a strict religious disciplinarian who always went into battle with an open Bible propped up on his saddle. Only that morning he'd shouted at a man smoking his pipe that he was a disgrace to the regiment and should stop smoking at once. When Cardigan's order came, Shewell didn't hesitate. The Light Brigade formed up in three lines, Cardigan in the lead, Shewell's 8th Hussars at the rear. 'We advanced at a trot,' Seager wrote home the following day, 'and soon came under fire . . . from cannon and rifles' on both sides. 'The fire was tremendous, shells bursting among us, cannon balls tearing the earth up and rifle balls coming like hail. Still on we went, never altering our pace or breaking up in the least, except that our men and horses were gradually knocked over.' Seager and what was left of his squadron swerved from side to side to avoid the dead and wounded. 'Many of the poor fellows looked piteously at us for assistance but on we went. We had received our orders and neither shot nor shell, although it was cutting us down in every direction, would prevent our obeying them.'

The full horror of what Lucan and Cardigan had ordered their men into was now hideously clear. Six hundred horsemen had lined up at the valley's mouth only moments earlier, their six regiments a blaze of colour, the riders confident and eager to fight. Less than two or three minutes later the lethal effect of fire from three directions at sometimes point-blank range left scores of troopers dead and wounded. Many more were thrown off their wounded and dying horses; blood and swirling dust were everywhere.

Seager saw brother officers fall one after the other. 'Poor Fitzgibbon was shot through the body and fell . . . Clowes's horse was shot under him and the last seen of him he was walking towards where he started from and we suppose he was taken prisoner or killed . . . Clutterbuck, who was on my left, got wounded in the right foot . . . and Tomkinson who commanded the squadron had his horse shot under him.' But all the men still mounted rode straight on. None thought to try to escape. As William Howard Russell of *The Times* newspaper reported: 'They swept proudly past, glittering in the morning sun in all the pride and splendour of war. We could hardly believe the evidence of our senses. Surely that handful of men were not going to charge an army in position? Alas! It was but too true – their desperate valour knew no bounds, and far indeed was it removed from its so-called better part – discretion.'

As the charge reached the Russian lines, with men and officers falling all around him, Seager took command of the squadron and moved to the front. His horse Malta was hit by a ball through the neck just above the windpipe but 'went bravely on'. A large body of Russian lancers then appeared in the rear and Seager shouted to the colonel to allow his men to wheel around to face this new threat. 'They were three deep with lances levelled,' Seager recalled. 'I parried the first fellow's lance, the one behind him I cut over the head, which no doubt he will remember for some time, and as I was recovering my sword I found the third fellow making a tremendous point at my body.' Seager managed to stop the point of the lance with the hilt of his sword. The lance's point penetrated the metal bars of the hilt and grazed the skin on top of his second finger. It then passed right through his little finger between the second and top joint. Seager wrote that he'd probably be officially listed as 'wounded' but actually: 'I have only got a slight scratch that might look interesting in a drawing room.' He felt very little pain and managed to write the letter home in spite of his wound.

Once he'd passed through the Russian lancers, through the gunfire and smoke, Seager spotted his surviving comrades including his commanding officer retreating towards the British lines. He tried not to bunch up with them: 'We had to go through the fire in a scattered manner so as not to give them a chance of killing us.' He also saw a larger body of lancers coming up on his left to cut him off, so 'I put Malta to her speed and she soon got out of their reach, but the shot and rifle balls flew in great quantities, shells bursting just over my head with an awful crash.'

Seager finally found his way back safely out of reach of the fire, and counted the cost. 'That any of the Light Cavalry Brigade returned through the cross fire kept upon us was through the great providence of God to whom I am grateful more than I can express.' Nearly half the 600 men who charged with the Light Brigade were killed, wounded or taken prisoner. No fewer than 355 horses were killed or had to be put down afterwards.

Seager's unit won much praise for wheeling to attack the lancers, although he was upset that his colonel got the credit for it. Seager himself had after all persuaded him to give the order. But he wrote that everyone was saying the charge of the Light Brigade was 'a most gallant exploit . . . never excelled in history'. '*C'est magnifique mais ce n'est pas la guerre*,' commented a French commanding general in the Crimea, Pierre Bosquet. Nevertheless what Seager and his comrades did soon made them a legend. Little more than a month after the event Tennyson published his immortal lines 'Half a league, half a league, half a league onward . . . ', and the reputation of the British cavalry received a much needed boost. Nearly half a century earlier the Duke of Wellington had lambasted his horsemen for their debatable contribution to his victories in the Napoleonic Wars. The Crimea restored the cavalry to the pantheon of heroes.

The argument rumbles on over who precisely was responsible for the order that resulted in such a tragic loss of life. No one was able later to cross-question Captain Nolan, the man who delivered it, since he was blown to pieces in the first moments of the charge. Some suggest that he saw Cardigan heading in a direction other than the one he, Nolan, had intended, and that he raced frantically across the front of the charge in order to try to abort it. However it happened, the suicidal onslaught was utter lunacy and a shameful indictment of the poor quality of British military leadership. Thanks to William Howard Russell, one of the first war correspondents, and others such as Florence Nightingale the British public became aware of the serious deficiencies in their armed forces.

Edward Seager echoed the general disenchantment with the army leadership. Of Lord Cardigan he wrote: 'If pomposity and bluster are the requisites for command, he is the man.' Seager credited him with going up to the Russian guns 'gallantly enough', but accused him of cowardice when on 'finding it no joke he bolted and left his brigade to get back the best way they could.' Cardigan left the

Crimea before Christmas pleading ill health. As for Lucan, 'he has not the head for the command, and if we do not get someone with their wits about him we shall someday have another Balaclava.' Raglan, the commander-in-chief, agreed: 'You have lost the Light Brigade,' he told Lucan who was sent home in early 1855. A rumour went around that Cardigan would be recalled to the Crimea to replace Lucan, and Seager observed: 'If he does, we all feel that having lost one Cavalry division, we shall soon lose another.' In the event Cardigan did not return and the war in the Crimea wore on – increasingly unpopular – for another year until the final defeat of the Russians at the collapse of Sebastopol in early 1856.

By January 1855 – with a whole year of the war still to run – Seager was in despair about what further contribution the cavalry could make. 'The Light cavalry, I am sorry to say, is no longer in being. In another month, I don't suppose there will be another horse alive in the Brigade.' In early 1855 he was posted back to Turkey to supervise a new cavalry depot there. Here he met the famous Florence Nightingale and was not impressed: 'I cannot say that I have much respect for women that step far out of their places and stations in this life,' he wrote to his brother. By the time Seager went home, the widely praised nurse had become an established figure in the Crimean War medical service 'not much to her credit or feminine feelings'. Seager also claims that Nightingale was unpopular with many others during the Crimean campaign: 'The officers here don't feel at all inclined to subscribe to the Nightingale fund. She would never allow a nurse to attend an officer, let him be ever so ill.' Seager did go on to admit that he knew nothing 'from my own observations'.

At the end of his year helping to rebuild the cavalry he proudly proclaimed in a letter to his brother that it was 'now in first rate condition. The cavalry here are all very strong.' But thankfully they were never required to fight any new Balaclavas, and Seager returned home to Emily and his children in the summer of 1856. He retired as a lieutenant general in 1881 and died two years later at the age of seventy.

2

Joshua Chamberlain

Professor and American Civil War Commander, 1862–5

SIX MONTHS BEFORE the First World War, on 24 February 1914, one of the world's finest old soldiers lay back on his pillow and died peacefully at the age of eighty-five. With his shock of white hair and billowing white moustache, Joshua Chamberlain was a real American hero. He entered the American Civil War as an enthusiastic volunteer on the Union government's side in 1862 and left it, a general, just three years later. During that short spell he escaped death more times and achieved more fame for his outstanding courage than many who spent a lifetime under arms. We find him particularly fascinating because nothing in his previous life gave any indication that he'd be an exceptional soldier. It left us wondering if great commanders are born and not made.

For most of his life Chamberlain was an academic, a college professor, lawyer and state governor. But a year into the American Civil War the thirty-three-year-old professor of modern languages took the lonely decision, without consulting his wife Fanny, whom he adored, to drop his comfortable life at Bowdoin College in

Brunswick, Maine, and throw himself into the fight for the Union. After watching so many of his students leaving to fight for a cause he passionately believed in, he opted to follow their example and joined the Union army. He was appalled at the breakaway by the states of the Southern Confederacy a year earlier. It was 'treachery', and it was, he said, his duty 'to sacrifice . . . dearest personal interests, to rescue our Country from Desolation'.

On 18 August 1862, dressed in a smart newly tailored uniform of a lieutenant colonel of infantry, he said goodbye to his weeping wife and four young children and took himself off to a nearby army training camp. He made a striking figure, at nearly six feet tall with a high forehead, sparkling blue eyes and a huge bristly moustache. For a man who was utterly unversed in military training, it was a baptism of fire, but his intellectual energy set him delving deep into all sorts of tactical manuals. Two weeks later when his brand-new regiment, the 20th Maine, was ordered into the war he felt ready for anything. The new volunteer found himself second-in-command to a battle-hardened colonel six years younger than himself. They made a good team.

What strikes us about Joshua Chamberlain's short military career as well as his courage is the startling number of times he escaped being killed. His first brush with death was on the first day he heard a shot fired, when the university professor so new to war found himself in the thick of battle. The 20th Maine joined the Union army of General George McClellan and on 20 September the regiment took part in McClellan's perfunctory effort to expel the Confederate force of General Robert E. Lee, which had advanced into Union territory on the Potomac River. Chamberlain was shepherding his men across the river, mounted on his horse in midstream facing strong gunfire from the far bank. 'The balls whistled pretty thick around me,' he wrote to his wife later, 'and splashed on all sides but didn't touch.' His horse however didn't escape: it collapsed wounded in the water and Chamberlain was propelled head first into the river. He struggled through the stream and sloshed to the bank unhurt. He knew Fanny would be appalled at the news of his narrow escape on his first day of battle, and as he rode south into Virginia with the regiment he wrote to her: 'Most likely I shall be hit somewhere at sometime, but all my times are in His hands, and I cannot die without

His appointing . . . I long to see you – to rush in and have a good frolic with the children, and a sweet sit-down with you in the study.'

For the advance into Virginia the Union's President, Abraham Lincoln, sacked the dilatory McClellan and appointed Ambrose Burnside. He was to prove a disappointment too and Chamberlain spent the winter fighting desperate but indecisive battles in Confederate territory, tempting fate as was becoming his habit, always by leading his men from the front. He had another close call when he allowed himself to stray in among the pickets of the enemy and was very nearly taken prisoner. He himself attributed his ill-judged detour to 'the rashness of youth'.

His next escapade cost him another horse. He allowed himself to be carried away by the excitement of an advance on a Confederate position, waving his sabre over his head with one hand and brandishing his pistol in the other. An enemy bullet hit his horse in the brain. Chamberlain was thrown off and hit the ground hard. He had to make his way back to the rear but with no more injuries than several bruises.

Chamberlain's big break came when he was promoted to lead the 20th Maine. Within a year of joining the army he had achieved his dream – commanding his own regiment just as Robert E. Lee, the brilliant Confederate commander-in-chief, made another audacious thrust across the Potomac into Union territory in Maryland. On 7 June luck, which had been in his favour so far, dealt Chamberlain a potentially fatal blow. He was leading his men back in extreme heat in what he called 'toilsome and hurried marches' and like many of his soldiers he went down with severe sunstroke. He later said he 'came near dying'. But by 29 June he was back at the head of his men in time to lead them at the critical Battle of Gettysburg. It was to be his finest hour.

The Union army – retreating into Maryland before Lee's deepest ever advance northwards – took up position on a long escarpment south of the town of Gettysburg called Cemetery Ridge facing the Confederate army on Seminary Ridge opposite. Chamberlain's 20th Maine found themselves on the extreme southern end of the Union line where the ridge fell away into a valley. Lincoln had made yet another change in his army commander and General George Meade was now in charge. His imperative was to hold Gettysburg and

Cemetery Ridge. If Meade was forced to retreat, the road to Washington would lie wide open to Lee. Victory for Lee at Gettysburg would be such a strategic triumph that it could prompt countries like Britain and France to recognise the Confederacy. It might well be the end for Abraham Lincoln's Union. The fight for Cemetery Ridge, now known as the Battle of Gettysburg, would prove the most decisive of the Civil War.

As Chamberlain moved his men forward on to the knoll named Little Round Top at the southern end of the ridge it quickly became clear to him that he held the most crucial ground in the entire Union line. If the Confederate troops approaching the ridge could push his men off the high ground or outflank them by fighting their way into the valley on Chamberlain's left, the Union army would be exposed to attack from behind. As the sun rose on 2 July 1863, Chamberlain received a chilling message from Meade: 'You are to inflict instant death on any who do not do their duty this day.' With the Confederate army massing for a head-on attack the vulnerability of Meade's extreme left was now apparent. But Chamberlain had only one regiment of fewer than 400 men to cover the whole of Little Round Top. The rest of his brigade was fully occupied defending further along the ridge to the north on his right.

General Lee and his formidable Confederate subordinate General James Longstreet agreed that Little Round Top had to be captured. They directed two Alabama regiments at the extreme end of the Union's ridge. Longstreet ordered Colonel William Oates of the 15th Alabama to conduct the key assault on the hill. Oates was five years younger than Chamberlain and also a former schoolteacher but two years of brutal war had made him an experienced fighter. As Oates and his men emerged from cover at the bottom of the hill, Chamberlain had to extend his line to the left and then bend it back in a great curve to stop the enemy attacking his rear. His regiment managed this shift of position very smartly. It was, he said later, 'so admirably executed by my men that our fire was not materially slackened in front while the left was taking its new position.'

Within two minutes the enemy was advancing upon Chamberlain's position firing rapidly – the first of several assaults. Chamberlain believed – confirmed by Oates – that his enemies were 'astonished' that they faced such a solid front. The Confederates trudged up the hill in one wave after another and attacked the 20th Maine. 'Our volleys,' reported Chamberlain, 'were so steady and telling that the enemy were checked here and broken.' The struggle went on for an hour and a half and 'was desperate in the extreme'. Chamberlain counted four times that the Confederates threw themselves at his men: each time they came as close as ten yards. Each time they were repulsed. 'The edge of the conflict swayed to and fro with wild whirlpools and eddies.' In the chaos of the struggle Chamberlain recalled sending back one man who had been shot across the forehead to 'die in peace'; but within minutes the man returned with a bandaged head fighting again with the fearlessness of 'those that shall see death no more'.

Chamberlain himself had several close calls. A Confederate soldier spotted that he was clearly an important officer and drew a bead on him. But he found he couldn't pull the trigger. Something stopped him and after the war he wrote to Chamberlain to say how happy he was that he hadn't fired. Then, said Chamberlain, came 'the last and most desperate assault' and his men found themselves running out of ammunition. The 20th Maine's fire slackened and officers rushed up to their commander shouting that the regiment was 'annihilated' and that men were looking for a way to pull back. Wounded and dying lay everywhere. Chamberlain himself was hit twice: his instep was cut by a shell fragment and his left thigh was badly bruised by a musket ball that struck his scabbard. Facing imminent defeat, he now acted like a man possessed. Instead of pulling back he shouted one word at the top of his voice: 'Bayonet!' 'It caught like fire,' he said later, 'and swept along the ranks.' Then the order 'Forward, at a run!' and the whole regiment, or what was left of it, charged down the hill. 'My men,' said Chamberlain, 'went down upon the enemy with a wild shout.' Somehow he managed to wheel his curved front rank back into a straight line to meet the enemy head-on.

The effect of this audacious bayonet charge was miraculous. The enemy turned and ran. One young Confederate officer tried to stem the flood. 'I was confronted by an officer,' wrote Chamberlain, 'who fired one shot of his revolver at my head within six feet of me.' This time Chamberlain's luck held: the pistol shot just missed his head. 'When, in an instant, the point of my sabre was at his throat, he quickly presented me with both his pistol and his sword, which I have preserved as memorials of my narrow escape.'

General Robert E. Lee's other assaults on the Union line failed too and Chamberlain's defence of Little Round Top and his well-timed switch from defence to attack made him a hero. People were soon crediting him with changing the course of the Civil War. Chamberlain, who was as vain as he was brave, wrote to Fanny: 'I received the personal thanks of all my commanders. When . . . I rode off from the field . . . the Brigade Commander took me by the hand, and said "Col. C. your gallantry was magnificent, and your coolness and skill saved us."'

Chamberlain had survived the Battle of Gettysburg with minor wounds and a close shave from the pistol shot, and he was delighted at being promoted to command a brigade at the end of the summer. But in the autumn of 1863 he had another lucky escape. On 7 November he was again thrown off his horse when its leg was pierced by a rifle ball in a fight near Rappahannock Station in Virginia. He managed to scramble to his feet and continue in command but six days later, after sleeping out in very cold weather, he collapsed with a fever and was taken unconscious to hospital in Washington. Doctors pronounced him 'dangerously ill'. A whole six months passed and it was not until May 1864 that Chamberlain managed to persuade his commanders that he was fit enough to be back on the battlefield.

He rejoined the Union army when its new commander General Ulysses S. Grant was locked in a bitterly contested but inconclusive battle with Lee's Confederates at Spotsylvania, forty miles south of Washington. Chamberlain relished being back in the firing line. Within four days of retaking command he wrote: 'The artillery fire is very hot now. Shells are bursting over us every second. The Brigade is losing men fast.' For the next four weeks he was in his element, engaging the enemy in one encounter after another, constantly exposing himself to hostile fire against the advice of many of his subordinates and earning the promise of promotion to the rank of brigadier general from his divisional commander.

By mid-June Grant had advanced his Union army south to the town of Petersburg beyond the rebel Confederate capital, Richmond. If Petersburg could be captured Richmond would surely follow. And

Chamberlain's men were the spearhead. On 18 June he led an attack on a battery of guns. Advancing, he was knocked off his horse. So was a soldier carrying the colours. Chamberlain promptly got to his feet, seized the fallen banner and charged forward waving it high in the air. The position was taken. It was, as Chamberlain recalled later, a 'sharp and hot, but short, encounter . . . we carried everything'.

A little later that day he and his brigade were ordered to lead the army's attack on the main Confederate line. It was a huge challenge. The enemy were well placed, commanding a wide field of fire. Chamberlain went up and down his line preparing and encouraging his men. One soldier remembered him making a rousing speech to his unit: 'Comrades, we have now before us a great duty for our country to perform, and who knows but on the way in which we acquit ourselves in this perilous undertaking may depend the ultimate success of the preservation of our grand republic.' Then at exactly 3 p.m., shouting, 'Come now, my boys, follow me,' he advanced on foot at the head of the brigade.

This was the moment when Chamberlain's luck finally appeared to run out. He was a conspicuous target, leading his men and waving his sabre above his head. They began at a quick march, then a double-quick march, and then broke into a run for the final charge. A mass of fire hit them – muskets, rifles, shellfire and canister knocking men down in scores. 'My staff were scattered,' recalled Chamberlain, 'my flag bearer shot dead, my own horse down. To cheer and guide the men, where no voice could be heard nor rank distinguished, I picked up the flag and bore it aloft.' Suddenly a ball from a rifle ricocheted from a rock and smashed upwards into Chamberlain's right hip tearing its way through his groin and lower abdomen, finally lodging in his left hip. He remembered feeling excruciating pain first in his back then further down. Blood poured down his legs from his hips. Determined not to fall and dishearten his men, he slammed the point of his sabre into the ground and struggled to stand as long as he could.

Once his men had rushed by, he collapsed first to one knee and then in a heap in the mud. He believed he was mortally wounded and it was not until well after nightfall that he was taken to a hospital three miles away. A surgeon offered to move a wounded corporal from an amputation table to make way for the brigade commander but

Chamberlain told him: 'Lay me to one side; I am all right.' The doctor ignored him and after a thorough examination pronounced him a hopeless case. In those days soldiers with torn internal vessels were usually left to die. Fortunately Chamberlain's brother Tom appeared with two very skilled surgeons. They managed to perform a delicate operation to staunch the flow of blood, repair what they could and extract the ball. Chamberlain was in excruciating pain but he constantly encouraged the two surgeons to persevere. It was touch and go.

Chamberlain's divisional commander came by, and, convinced that he would not last the day, sent a message to General Grant to request immediate promotion for the colonel he believed was dying. Back came the message: go ahead. The newly promoted brigadier general wrote to Fanny: 'My darling wife I am lying mortally wounded the doctors think. God bless and keep and comfort you, precious one, you have been a precious wife to me. To know and love you makes life and death beautiful.'

But by now Chamberlain had a knack for cheating death. Against all the odds he recovered. After several months of treatment, he returned to the battlefront in March 1865.

The war was in its final weeks. The grinding down of Confederate strength in battle after battle combined with the North's superior industrial power left General Lee's army beleaguered on the Appomattox River west of Richmond. Chamberlain, his reputation for leadership, gallantry and survival now unassailable, was welcomed back to command a brigade for the last few weeks of bitter fighting. On 29 March he found himself in the forefront of an assault up the Quaker Road. He was just about to reach the enemy's line at the head of a charge when – perhaps in a momentary realisation that he was being reckless – he tried to slow his horse down. The horse reared its head and a sharpshooter's bullet hit the horse's neck and then struck Chamberlain – so he claimed – near the heart and passed 'around two ribs so as to come out at the back seam of my coat'. It also hit 'a leather case of field orders and a brass-mounted hand-mirror in my breast pocket'. Some of his comrades-in-arms, recognising his gift for exaggeration and flowery language, wondered if he was wounded at all. And it certainly amazes us that the brigade commander managed to throw himself into the final week of battle with remarkable energy for a man just pierced by a bullet from front to back.

On 9 April Chamberlain was leading his men towards Appomattox when an officer waving a white towel approached from the Confederate line. As the man approached, Chamberlain recalled that his own mood was 'so whimsically sensitive that I could even smile at the material of the flag – wondering where in either army was found a towel, and one so white. But it bore an almighty message . . .' Chamberlain was witnessing the surrender of the Confederate army. It wasn't long before he saw the dignified figure of Robert E. Lee riding towards the Union lines with one attendant to meet Ulysses S. Grant in the Appomattox courthouse and end four years of unimaginable bloodshed. Chamberlain described Lee as 'a commanding form, superbly mounted, richly accoutred; of imposing bearing, noble countenance, with an expression of deep sadness overmastered by deeper strength'.

But Chamberlain's most treasured moment was still to come. Grant asked him to command the ceremony in which the Confederate forces would formally surrender their weapons and colours. Three days later he paraded his men to witness the downcast ranks of the defeated Confederate army marching past. 'The momentous meaning of this occasion impressed me deeply,' he recalled. 'I resolved to mark it by some token of recognition . . . Was not such manhood to be welcomed back into a Union so tested and assured?' In a gesture that was to win him both acclaim and disapproval, Chamberlain ordered his buglers to sound and his soldiers to raise their weapons to the position of the traditional salute. And the surrendering Confederate general responded by ordering his men to return the compliment.

For the rest of his long life – he became Governor of Maine and lived until the eve of the First World War in 1914 – Chamberlain remained proud of what he had done that day. And one old Southern veteran remarked that the reunion of the two sides in the Civil War 'began with that order to present arms'.

For us Joshua Chamberlain is one of the most captivating characters in military history. A remarkable mixture of modesty, conceit and, above all, astonishing bravery, he deserves all the praise that was showered on him for the part he played in America's bitterest internal conflict both for his treatment of his men and his conduct in battle.

3

Private Johnson Beharry VC

Firefight in Iraq, 2004

Irene Beharry kept telling herself not to cry. She was a nurse and had seen some terribly damaged people before, but the man lying on the trolley in front of her was her nephew, Johnson. His head was the size of a football. He had lacerations and bruises all over his face. His puffed and swollen eyes seeped fluid.

'Once under general anaesthetic,' the neurosurgeon told her, 'Johnson will be placed on the operating table and an incision will be made across the top of his scalp from one ear to the other and the coronal flap pulled down in front of his face to expose the compound fracture. The surgeons in Kuwait did a good job but we now have to spend eight hours repairing the fracture and rebuilding the airways around his nose.'

He reassured her that Beharry's chances of survival were good. 'Your nephew is young and strong. I have been told about some of the things he did in Iraq. If anyone can pull through what lies ahead, it's Johnson.'

Irene took her nephew's hand and he opened his eyes and looked at her. 'Auntie I,' he said sleepily, 'you all right?'

She told him she was fine and he would be too.

'Don't go blaming yourself, Aunt I. Sometimes I know you think this is all your fault.'

She felt the tears coming again. 'That's true,' she said, 'I do.'

'You gave me another life, Auntie I. Joining the army was my decision. I don't regret it. I did what I wanted to do.'

It was Johnson Beharry's Aunt Irene who was largely responsible for encouraging him to leave his native Grenada and try his luck in Britain. That was back in 1999 when he made the life-changing decision to leave his childhood home on the small West Indian island and fly to London. It wasn't just prompted by the enthusiastic response he got from his Aunt Irene and Uncle Raymond in London when he suggested it. It was also his sense that he could do better than scrape a living out of the grinding poverty of Grenada where he'd been born to a drunken bullying father and his long-suffering mother. The person Johnson loved more than anyone else was his grandmother, his 'Nan', who used to sit him on her lap and ask him what he wanted to do with his life. She told him he could do anything he wanted. 'I want to be a driver,' Johnson remembered telling her, 'a racing driver.' He used to dream of driving a brand-new Porsche 911 Turbo convertible. He'd make it from Diego Piece in the north of the island to the capital St George's in under twenty minutes. 'I'm the king of the road,' he'd say to himself, 'I have 300 brake horsepower in the compartment behind me.' In the real world he found himself working as a motor mechanic and earning what he called 'real money' as a painter and decorator. He met and married a young lady called Lynthia, whose grandfather fought for Britain in the Second World War. 'Any chance you have to travel,' said her grandfather, 'you have to seize it. That is what the army gave me – a chance to see something of the world.'

Beharry was twenty when he arrived in Britain in 1999. His Auntie Irene persuaded him to stay in the UK. 'England may not be perfect,' she told him, 'but it has allowed me to become something.' He didn't need much convincing. One day he was reading a newspaper in the Tube and spotted a recruiting ad for the British army. He was immediately captivated by the idea of earning some decent money and getting himself a British passport. There would be downsides. 'The army is seriously racist,' said one of his friends. 'You won't last a minute.' Auntie Irene was deeply anxious when Beharry said he was off to Catterick in Yorkshire to train with the Prince of Wales's Royal Regiment (PWRR), affectionately known as 'the Tigers'. 'It will break your mother's heart,' she said, 'all that danger. So much fighting in the world. Why put yourself in the firing line?' Between them they agreed not to tell Johnson's family in Grenada for the moment.

The battalion Beharry joined in 1PWRR consisted of around 600 men: he found himself in a platoon of thirty men. His platoon operated out of four Warriors – armoured infantry fighting vehicles – each armed with a 30mm cannon. It was a sort of armoured taxi that could carry up to seven men. It was proof against small arms fire but not against larger armour-piercing, rocket-propelled missiles. Beharry made no secret of the fact that he'd like to be a driver. He could soon rattle off a Warrior's specifications. Its 550 HP Rolls-Royce engine gave it a maximum speed of over fifty miles an hour. The commander and gunner sat in an electrically operated turret controlling the Rarden cannon and the Hughes 7.62mm chain gun. The driver sat encased below. He normally drove with his hatch open but if forced to close it under heavy fire he could only see out through a periscope. A Warrior in motion made such a thunderous noise that its crewmen were forced to communicate with each other by radio.

Beharry soon found himself teamed up with another West Indian, his gunner, named Samuels, who came from Jamaica. Everyone called him Sammy. Sammy called Beharry 'Paki' because his light-coloured skin was more brown than black. There was a lot of racism in the army, but Johnson learned to cope. He was not the only non-white, not by far. At one time there were Sammy and two other Grenadians in his crew. One particular sergeant trainer at Tidworth, Sergeant Chris Brome – they called him Broomstick – referred to them as 'the four bloody stooges'. He was particularly amused at Johnson's name: 'Beharry,' he jibbed, 'Grenadian name. Originally from somewhere near the Taj bloody Mahal.'

By 2004 Johnson Beharry was a fully qualified driver and his Warrior, dubbed Whisky Two Zero (W20), was as precious to him as if it were his home. He knew every nut and bolt, every link in its tracks. He'd proved himself capable during a stint in Kosovo, where British forces had been part of the NATO intervention that pushed the Serbs out of the tiny Balkan state.

Now his battalion was off to Iraq as part of the occupying force after the American and British invasion. While the Americans were in the turbulent parts of central and northern Iraq the British were headed for the south. 'We're all excited,' Beharry remembered thinking at the time. 'The war is over. The Americans may be getting a hard time but the papers are full of British soldiers strolling around Basra in their berets, talking to the locals and handing out sweets to children.'

Little did he know. It was true that the British-American invasion a year earlier had been a military success. Saddam Hussein had been removed from power and his regime was no longer a threat to anyone. But the political impact of the occupation of Iraq was disastrous. Most of the rest of the world judged the British and Americans were wrong to target it. The much trumpeted pretext that Saddam's Iraq was awash with weapons of mass destruction turned out to be a myth. Far from celebrating their liberation from the tyranny of Saddam, many Iraqis resented the foreign intrusion. In the chaos and anarchy of the invasion's mismanaged aftermath, the Americans – and soon the British – found themselves the targets of all sorts of militia groups including the dreaded Al Qaida.

By the time Johnson Beharry drove his Warrior off the ship into Iraq in early 2004, the place had turned very hostile. He and his team soon found themselves in Al Amarah, 100 miles north of Basra where they were opposed by the fanatical Shia Mahdi army. The British had shut down a newspaper belonging to the Mahdi army's leader, Muqtada al-Sadr. 'He's an angry man right now,' Beharry's new CO told the regiment. 'We do know the Mahdi army is well armed and we have to assume they are capable of causing us trouble if they want to.' Whisky Two Zero's crew was now Beharry, the driver, Sammy, the gunner, Jamaican Kevin Campbell and the platoon commander 2nd Lieutenant Richard Deane. Deane was a former bank employee from Northern Ireland. 'They tell me you're the best driver in the battalion,' he told Beharry. 'They also told me not to tell you, because it'll swell your head. And,' said Deane, looking at Sammy, 'this I take it is your mate Samuels. I hear that you two are as thick as thieves.'

It wasn't long before Whisky Two Zero's tank park was being shelled by the militants and the decision was taken to go into Al Amarah's housing district and snatch around a dozen leaders of the protest movement. 'At long last,' wrote Beharry later, 'we're doing something about the shit we've been getting since we arrived.' It was 1 May 2004.

Within minutes of hitting the road Beharry and his crew knew they were in a deadly war zone. The hot dusty streets were mostly deserted but suddenly a makeshift barrier appeared ahead. Beharry, driving, heard Sammy say, 'Shit!'

'What?' asked Deane.

'Left-hand side. There's a kid across the street. Eleven, maybe twelve years old. He's holding what looks like a fucking RPG' (a rocket-propelled grenade, lethal if it strikes an armoured vehicle at the right angle).

As Beharry turned to look, there was a massive explosion and the Warrior shook. He called out to the others, but there was no reply. Whisky Two Zero stood there like a double-decker bus in the glare of the street lights. RPGs tend to come in threes. Instinct made Beharry hit the accelerator but he was an instant too late. A second explosion hurled Whisky Two Zero across the road. Beharry found himself driving straight at the barrier. By this time smoke was billowing through from the back of the Warrior into the driver's compartment.

'Drive, Paki, for fuck's sake, drive!' shouted Sammy. 'There's more of them lining up with RPGs.'

Beharry spotted gunmen shooting at them from the rooftops. RPGs can bore through ten centimetres of armour. Beharry didn't lose a second. The barrier gave way to the Warrior's pressure. There was a lot of shouting from the back of the vehicle. Someone called out for Deane, the commander.

'Stop calling the boss,' shouted Sammy. 'The boss is dead! He's lying on the floor of the turret.'

A man ran out and sprayed the Warrior with his AK 47. Beharry, desperately trying to control the Warrior, shouted at Sammy to take him out with his machine gun.

'I've got a stoppage!' Sammy screamed back.

The assailant went on firing and then darted into an alleyway, to be replaced by a man carrying an RPG. Beharry glimpsed him and managed to slam his hatch shut. Just in time. An ear-splitting explosion tore the handle out of his grip. There were screams behind him and no response when he shouted back for Sammy.

With the other Warriors in the platoon following along behind, Beharry weaved his way down the street, took a quick turning down a side street then pulled up beside another Warrior, his company commander's. There was a fierce firefight going on but Beharry realised he had to try to transfer Sammy and what was left of Deane to another vehicle. He pulled himself out of the driver's hatch and looked down into the turret. Deane was on the floor and Sammy was hunched up clutching his sides. Dodging bullets, somehow Beharry managed to drag both out. He heaved one into the back of the commander's vehicle and the other into a next-door Warrior. Then he zigzagged back to his crippled and smoking Whisky Two Zero to disable the Rarden cannon so that the militants couldn't use it. Moments later he staggered into the back of another Warrior and passed out.

When he woke up in hospital his only complaint was heat exhaustion. His company commander told him that what he had done was outstanding. 'And Richard Deane is going to be just fine, thanks to you.' Sammy was recovering fast too.

The company sergeant major walked in with Beharry's helmet. It had a hole in the top of it and a three-inch-long groove. A 7.62mm round was still in the lining. 'That's a pretty bloody close shave you had there, Beharry.' Johnson remembered the moment when he'd shouted at Sammy to use the machine gun and something had slammed into his head, knocking it back against the hatch. 'The helmet's yours. Something to show the grandkids.'

Six weeks later, with the situation further deteriorating in Amarah, Johnson Beharry's lucky streak ended. At 1.30 a.m. on 11 June he and his Whisky Two Zero, now fully repaired, were on mortar 'rat-trapping' duty. When mortar bombs were fired at the British camp and the source could be identified, Warriors were sent out to track down the bombers. This time it was Richard Deane's turn. He gave Beharry directions to the grid reference he'd been passed and they drove off. Within twenty seconds the street lights switched off. 'They know we're coming,' Beharry said to himself. Navigation was almost impossible. There were night sights but the only way to be sure of where he was going was to open the hatch and look out. Beharry strained his eyes. He spotted a cat walking across the road. Then there was a flash off to his left. Deane shouted a warning and he heard Sammy yell his name. Beharry remembered another flash of light, a massive crash and then something slammed into his head. He couldn't see and his ears seemed to be blocked up.

But somehow he managed to hear Richard Deane shouting, 'Bee, can you hear me?'

'I hear you.'

'Get us the fuck out of here! Go, go, go!'

'Okay, boss.'

Instinctively Beharry managed to slip the gear into reverse and gun the accelerator. The Warrior shot backwards and eventually hit a wall and stopped.

The vehicle had actually been struck by several RPGs, one of them shattering the strut that holds the driver's hatch open. It exploded only a foot from Beharry's face. If he hadn't been so tall, which made him lean backwards in the driving seat, he would have been killed outright. Miraculously he survived and even more remarkably – in spite of suffering a severe brain injury – he was able to respond to Deane's desperate order to withdraw. Another Warrior drew up beside them and Beharry was very gently eased out of the hatch and into the back of the other vehicle. Richard Deane, badly injured in the face, was transferred too.

A Chinook helicopter sped Beharry to hospital in Kuwait. A surgeon explained to the regiment's commanding officer that the pressure wave from the exploding RPG had caused multiple fractures to his skull. He was in a coma and any further intervention would require specialised neurosurgical skills not available in Kuwait

Beharry's Aunt Irene and his wife, Lynthia, were flown out to Kuwait. Four days after the incident they entered the room where he was lying with tubes coming out of his nose and drips in his arms. He was still in a coma. He didn't recover consciousness for eight days after the explosion. Three days later he was flown back to Britain, to the neuroscience centre in Birmingham. He underwent an eight-hour critical operation and then woke up to find Auntie Irene and Lynthia looking down at him.

A month later, Johnson Beharry was admitted to Headley Court, the military rehabilitation centre, where they rebuilt the muscles he had damaged in his back while pulling Richard Deane out of his hatch after the first close encounter with the RPGs. Months later the regiment came home from Iraq, and Beharry, who had been told he was no longer fit enough to remain in the army, met Deane, Samuels and all his old comrades. The battalion's commanding officer told him that the battalion had suffered 850 attacks in its six months in Iraq and they had cost the regiment two dead and forty wounded including Beharry. Then the CO delivered spectacular news. Deane and Samuels were to receive the Military Cross for their bravery in Whisky Two Zero. Beharry was to get the Victoria Cross, the highest of all British decorations for courage. His was the first VC to be awarded since the Falklands War in 1982.

As the Queen pinned the VC on Beharry's chest, she said, 'It's been rather a long time since I've awarded one of these.'

Just over a year later Beharry returned to Grenada to a hero's welcome. Even his intemperate father embraced him and said how proud he was.

Sadly Beharry carried the scars of those terrifying days in Iraq into later life. In December 2008, beset by memories he found hard to bear, he attempted suicide. He drove his car at 100 mph into a lamppost. He was knocked unconscious but survived and candidly admitted that he intended to end his life. 'Everyone thinks I am a hero because I was awarded a VC. But I'm just a normal soldier who cannot get away from his demons.'

He later managed to put this behind him by launching in 2014 the Johnson Beharry VC Foundation – the JBVC. It's a charity that works to persuade youngsters to leave street gangs and develop other interests. Beharry's most precious memory today is of the small boy of six or seven wearing a torn T-shirt and no shoes who welcomed him on one of his visits home to Grenada with the words: 'Johnson Beharry! I want to be you.'

MESSENGERS

ONE OF THE greatest transformations in the nature of warfare is how news of troop deployment, enemy movement and victory or loss is transmitted. Military leaders in today's digital age will soon be able to don a virtual reality headset and explore the battlefield in real time. This instant access to information is a dramatic change from a not-so-distant time when news from the battlefield was hand-carried by messengers. Reports of these intrepid couriers date back to 490 BC when fleet-footed Pheidippides ran twenty-six miles from Marathon to Athens to announce the Greek victory over the Persians in the Battle of Marathon. After uttering the words, 'Joy to you, we've won,' the exhausted messenger allegedly collapsed and died. The legacy of his heroic dash lives on in today's twenty-six-mile marathon. In Roman times, communications and roads were designed to allow imperial forces to report and crush rumbling discontent.

Messages in the past often arrived too late. In January 1815 British and American forces fought a bloody battle at New Orleans two weeks after a peace between the two sides had been agreed in Europe. News of the Treaty of Ghent that ended this War of 1812 took nearly two months to travel by ship to America.

Commanders in the old days used every means to speed communications – from smoke signals to bonfire beacons. Animals played their part too: a messenger travelled fastest on horseback until the nineteenth century; carrier pigeons proved the best way of relaying messages from the besieged population of Paris in the Franco-Prussian War of 1870–1; dogs carried vital messages between trenches as late as the First World War.

As journalists we have always been curious about how news travels.

It was the invention of the telegraph in the mid-nineteenth century that greatly speeded up the delivery of information. One of our heroes is William Howard Russell whose pioneering telegraphic dispatches from the Crimean War revolutionised war reporting. Today's war correspondents can file reports and pictures twenty-four hours a day from a mobile phone. Civilians under fire share scenes from their living-room windows.

One of history's most infamous messengers was Adolf Hitler who was a regimental runner in the First World War. He was wounded twice and received Iron Crosses, second and first class for his pains. We have chosen more admirable messengers: a dashing British officer who, in 1815, raced to London with the first news of Wellington's victory over Napoleon at the Battle of Waterloo; and a humble housewife from Upper Canada who carried a crucial warning during the War of 1812. But we start with a lone American on horseback who raised an alarm in 1775 and would have been long forgotten had he not been immortalised in a poem by Henry Wadsworth Longfellow.

4

Paul Revere

Revolutionary Messenger, 1775

Listen, my children, and you shall hear
Of the midnight ride of Paul Revere.

HENRY WADSWORTH LONGFELLOW did much to create the legend that surrounds one of the heroes of the American War of Independence. One hundred years later, on the eve of the US Civil War, he was searching for inspiring Americans to write about. He stumbled across the little known story of Paul Revere. As every American schoolchild now learns, Revere rode through the quiet New England night to alert American patriots that their British enemies were on the march. Revere's ride on 18/19 April 1775 was the spark that helped ignite the American rebellion against the British Crown. But what few people know is that his involvement in the eight-year war that was to shatter Britain's hold on its American colonies went much deeper than that single dramatic midnight ride.

Sixteen months before Revere's famous ride, a group of men silently boarded three British ships anchored in Boston Harbour under cover of dark. Some were disguised as Mohawk Indians, their faces smeared black with coal dust and streaked with red warpaint. They wore wool blankets and carried tomahawks – a clear signal that they regarded themselves as citizens of America rather than Britain. On that memorable night, 16 December 1773, one of the malcontent leaders was Paul Revere, a thirty-nine-year-old silversmith and engraver. For the next three hours he and others smashed open 342 chests of tea and dumped it overboard. The Boston Tea Party was a violent protest against British taxes and laws which a growing number of Americans regarded as unfair and unacceptable. Within days of the audacious raid came a street ballad praising 'brave Revere . . . fighting freedom's cause' and urging:

> Rally Mohawks. Bring out your axes
> And tell King George we'll pay no taxes.

'Brave Revere' was a French immigrant's son who inherited the family's modest silver business in Boston when he was just nineteen. Married at twenty-one, and desperate for money to feed his fast-growing family, he started making false teeth out of ivory. He advertised 'persons so unfortunate as to lose their Fore-Teeth by Accident, and other ways, to their great detriment not only in looks but speaking may have them replaced with artificial ones.' (Years later he scored a forensic first, identifying a body by a false tooth he had installed.)

Hardworking and ambitious, Revere, like so many other American patriots, resented British rule, especially the punishing taxes the government in London imposed on its thirteen colonies. His antagonism began back in 1765. The mother country was deeply in debt after a costly seven-year war with France and the American colonies were an obvious source of income. Everything from sugar to stamps were taxed much to the fury of the colonists. 'Taxation without representation' became a rallying cry for people like Paul Revere, fed up with the British Parliament laying down the law. There were protest meetings in town halls and state legislatures. British goods were boycotted. A network of secret organisations called the Sons of Liberty sprang up, fiercely resisting British rule. Revere was an active member. A favourite tactic was to terrorise British tax collectors. One night, as two inspectors left a dinner at the Boston Concert Hall, Revere and several of his co-conspirators, faces blackened and wearing white nightcaps, jumped out and scared the living daylights out of them.

Revere was an artist too, and he used his skill as an engraver to protest. When two regiments of British soldiers arrived in Boston in 1768, to keep the peace, Revere published a picture that he entitled *That Insolent Parade*. Two years later, after British soldiers fired on an angry crowd in Boston and killed five men, Revere engraved and circulated a drawing called *Fruits of Arbitrary Power, or The Bloody Massacre Perpetuated in King Street*. It quickly became a symbol, an indictment of British tyranny.

But it was the Boston Tea Party in 1773 that marked a turning point in America's relationship with Britain. Revolutionary committees were set up throughout the colonies and the eager Revere became a trusty courier. He rode down the Atlantic east coast to New York and Philadelphia carrying news of unrest in Boston and bringing back reports of other planned protests. The protests grew after the British government imposed harsher measures, closing the port of Boston and taking away the right of Massachusetts politicians to elect members of the state's executive council. Most town meetings were forbidden and a new system of courts set up to impose British law. The British insisted these so-called Coercive Laws were necessary because the American colonies were 'in a distempered state of disturbance and opposition to the laws of the mother country'.

Dissident Americans called these new laws the Intolerable Acts. As more and more protest groups sprang up, Paul Revere and his horse became a familiar sight on roads from Boston to Philadelphia, galloping from village to village carrying revolutionary messages. In Boston he met with like-minded patriots at the Green Dragon Tavern: 'In the Fall of 1774 and Winter of 1775,' he later recalled, 'I was one of upwards of thirty, chiefly mechanics, who formed ourselves into a committee for the purpose of watching the movements of the British soldiers.' The growing rebel movement set up a system of 'alarm and muster', so that when the time came and the alarm went out volunteers would pick up their arms and rush to form units.

The British were well aware of Paul Revere's activities. When he raced to Portsmouth, New Hampshire, in December 1774 to check out rumours of British troops landing there, an official in Boston reported back to London. He wrote that Revere helped to assemble up to 500 armed men known as Minutemen because it took them just minutes to get ready to fight. They marched into Portsmouth, prepared to take on the Regulars, as British soldiers were called. When the rumours of a landing proved false, the Minutemen returned home but Revere knew he could count on them when the time came.

The British government massively underestimated the growing threat to their hold on the American colonies. With typical arrogance and complacency, ministers in London dismissed the growing number of rebels like Revere as 'a rude Rabble without plan, without concert, and without conduct'.

The showdown came in April 1775, when General Gage, the commander-in-chief of British forces in the New World, received orders from London to find and remove weapons hidden in Concord, a prosperous community twenty miles north-west of Boston. He was also told to arrest rebel leaders including Samuel Adams and John Hancock, members of the Massachusetts Provincial Congress, the provisional government set up a year earlier in defiance of the British. On 18 April Gage told his men to proceed 'with utmost expedition and secrecy to Concord, where you will seize and destroy . . . all military stores . . . But you will take care that the soldiers do not plunder the inhabitants or hurt private property.'

Secrecy was a problem in a city where every move of the British Regulars was being closely monitored and no one had his ear closer to the ground than Paul Revere. When the British march to Concord got under way on the evening of 18 April, he knew all about it.

A well-rehearsed plan went into effect. Revere had persuaded the sexton of the tallest church in Boston to place lanterns in its steeple if British troops headed out of the city. The church stood on the bank of the Charles River and its steeple could be seen for miles around. If one lantern appeared it meant the British were marching along the road. If there were two lanterns, they were moving across the river by boat. 'One, if by land, and two, if by sea', in the words of Longfellow.

Once he knew the British were crossing by boat, Revere leapt into action and raced to the sexton's house to tell him to light two lanterns. But when Revere peered through the sexton's window, he was horrified to see the man's parlour full of British officers playing cards. They were lodgers in the house. Revere stood in the garden wondering what to do next. Suddenly, the sexton slipped out of the shadows beside him. He had told his lodgers he was tired but instead of going to bed, he had quietly opened an upstairs window and jumped out. Revere passed on his message and left the sexton and two accomplices to climb up the 154 stairs to the church tower and light two lanterns.

Revere now had to warn the Americans in Lexington and Concord. Minutes later he was being rowed by two experienced boatmen across the Charles River. It was just as well they had decided to muffle the oars because a British warship unexpectedly loomed up in their path. The small rowing boat silently slipped by unseen and, once on dry land, Revere was met by compatriots who warned him British officers were patrolling the road. Unperturbed, Revere said, 'I went to git me a horse.' He was fortunate enough to be lent 'a very good horse' named Brown Beauty. According to Longfellow:

> The fate of a nation was riding that night;
> And the spark struck out by that steed, in his flight,
> Kindled the land into flame with its heat.

Revere soon spotted two British Regulars in the distance ahead and quickly turned around. The soldiers gave chase but Brown Beauty was too fast for them.

Revere was able to warn everyone on his route that the British were on the move. Legend has it that Revere shouted, 'The British are coming,' but eyewitnesses report he cried, 'The Regulars are coming out.' Once they heard Revere's shouts, other patriots who were ready to take messages jumped out of bed and on to their horses, spreading the word. The alarm and muster system so carefully organised months before worked perfectly.

When Revere arrived in Lexington around midnight, one disgruntled resident told him not to make so much noise 'because people are trying to sleep'. He replied, 'Noise! You'll have noise enough before long! The Regulars are coming out!'

He headed straight for the house where Adams and Hancock were staying. The pair had a heated discussion about whether the British were after weapons hidden at Concord, a few miles further on, or out to arrest the two rebel leaders. They finally agreed that since there was a large force of Regulars on its way – more than 700 in all – the target was most probably the stash of weapons in Concord.

Revere jumped back into his saddle to carry the warning on to Concord. He was joined by William Dawes, a tanner who had ridden to Lexington on a different route, and by a dashing young doctor named Samuel Prescott. In his official account of the ride, Revere tells us the doctor was 'returning from a lady friend's house at the awkward hour of 1 a.m.'.

Revere also recounts how he warned his companions, 'we might be stopped before we got to Concord . . . I likewise mentioned that we had better alarm all the inhabitants until we got to Concord.'

Halfway to Concord, their path was blocked by four British soldiers brandishing swords and pistols. One soldier shouted, 'God damn you! Stop! If you go an inch further you are a dead man.'

'We attempted to git through them but they kept coming before us,' wrote Revere.

Prescott, who knew the country very well, managed to escape. He headed straight to Concord and successfully sounded the alarm. Dr Dawes also got away but Revere found himself surrounded by ten Regulars. He was ordered to dismount at gunpoint. After being shouted at by the men who captured him, Revere was questioned by an officer he described as 'much a gentleman'.

'Sir,' the British officer said politely, 'may I crave your name?'

'My name is Revere.'

'What?' the officer exclaimed in surprise. 'Paul Revere!'

The 'gentleman', said Revere later, 'knew exactly who his prisoner was'.

When the officer questioned the messenger about what he was up to, he was astounded to discover that Revere was better informed about British troop movements than he was. Revere said that the

Regulars had left Boston by boat. 'Their boats had catched aground,' but the soldiers had eventually got ashore safely. Revere was desperate to keep the British away from Lexington where he remained convinced they planned to arrest Adams and Hancock. Hoping to frighten them off, he announced that he had already warned American patriots in Lexington that the Regulars were on their way. He told his captors that messengers had been sent out all over the surrounding country urging men to gather and fight in Lexington. A less sympathetic officer was summoned to hear Revere's tale: 'He clapped a pistol to my head and said he was going to ask me some questions, and if I did not tell the truth, he would blow my brains out.'

Revere was ordered back on his horse and escorted towards Lexington. Suddenly gunshots rang out. An officer asked Revere what it meant and he replied it was 'a signal to alarm the country'. Next, the town bell started ringing. Revere did his best to put the fear of God into the Regulars: 'The bell's a'ringing! The town's alarmed, and you're all dead men.'

The soldiers decided to retreat in order to alert the advancing British force to the dangers ahead. They felt there was nothing more to learn from Revere, so they let him go but confiscated his horse. It was now 3 a.m. and Revere walked in heavy riding boots through dark, swampy land back to Lexington. When he arrived at the house where Adams and Hancock were staying he was astonished to find that they were still sitting in the dining room arguing about what to do. Hancock, who was spoiling for a fight, had spent the night cleaning his gun and sword. Adams kept reminding him he was a politician not a warrior. War, Adams said, was 'not our business. We belong to the cabinet.'

Revere urged both men to leave Lexington immediately but according to Hancock's fiancée, 'It was not until break of day that Mr H. could be persuaded.'

After accompanying Adams and Hancock to the safety of a parsonage a few miles away, the exhausted Revere returned to Lexington and helped carry a trunk full of Hancock's papers to a safe place in nearby woods. 'We made haste & had to pass thro' our Militia who were on a green behind the Meeting House, to the number as I supposed, about 50 or 60.' From his hiding place in the woods, Revere watched 700 red-coated Regulars advance through early morning mist. When the Minutemen stood firm, the British commander shouted, 'Disperse, you damned rebels! You dogs, run!' A shot rang out. It is not clear who fired first, the British or the Americans, but soon all hell broke loose. There was, said Revere, 'a continual roar of musketry'.

Only one British soldier was wounded in the skirmish. Eight Lexington men died. Ten were wounded. Paul Revere was witnessing the first military engagement of the American War of Independence.

Realising they were hopelessly outnumbered, the American Minutemen finally dispersed. The British then marched to Concord, two hours away. Once there, they divided into smaller groups to search for rebel weapons. All they found were three cannon and some bullets. The rest of the rebel arms had been moved to other villages days before. While the British searched, more and more armed Minutemen were arriving in Concord. Up to 500 of them gathered in a field on one side of a river crossing known as North Bridge. They advanced on a smaller British unit on the other side of the bridge. One of the Regulars fired his gun. Others, believing they'd been ordered to fire, followed suit. The Americans fired back. The poet Ralph Waldo Emerson later described the first shot fired by the Americans at North Bridge as the 'shot heard round the world'.

Paul Revere may not have been in Concord to hear that shot but he and the other messengers had played a key role in igniting the American Revolution.

Outnumbered and outmanoeuvred in Concord, the British were forced to retreat back to Boston, harried all the way by American snipers hiding behind trees, hedges, houses and barns. The British lost 273 men. Ninety-five Americans died.

During the war with Britain that followed, Revere served in the Massachusetts Regiment but the bold messenger was a poor soldier. He was asked to resign from the army after being accused of disobeying orders and general incompetence during an amphibious assault at Penobscot Bay. True to form he requested a court martial to clear his name, arguing that he had acted to protect his men. He was lucky to be acquitted.

After the war he handed his silver business over to a son. Always open to new ideas, Revere learned how to cast metal, then started a foundry and made church bells. He also set up America's first successful sheet-copper mill and his products went on to adorn the top of the new Massachusetts State House. They were also used to protect the hull of the frigate USS *Constitution*, the first warship in the independent United States' navy. She was to sink several British ships during the War of 1812 and earned the name of 'Old Ironsides'. The ship can still be seen in Boston Harbour.

Paul Revere's final years were spent peacefully on his country estate where he was surrounded by his children and more than fifty grandchildren. He died at the grand old age of eighty-four in 1818.

Today the route of his famous midnight ride is a national park. Every year on 19 April the sound of a tolling bell rings out in Boston. It commemorates the renowned messenger who rode through the night to warn that the British were coming. The bell was made by Paul Revere.

5

Laura Secord

Niagara Heroine, 1812–13

ONE OF THE first snippets of history taught to schoolchildren in placid and peaceful southern Ontario is the story of Laura Secord, one of Canada's most unlikely wartime heroines. Generations of young Canadians have idolised the plucky housewife who walked twenty miles to warn British troops of an imminent attack by the Americans in the War of 1812. It was one of the few times that Canada was invaded. We've always found it an intriguing yarn even though it occasionally gets embellished in the telling. One of the most popular myths is that Secord took a cow and a milk pail through American lines and persuaded US sentries to let her pass because she said her animal needed milking. No one has found any record of a cow or a pail but there is plenty of proof that the thirty-seven-year-old mother of five risked all to carry a message that saved many British and Canadian lives and helped ensure an independent Canada.

Secord and her family had the misfortune to find themselves living in Queenston, a small town in the Niagara Peninsula in the middle of a war. An easy crossing point into Canada at the west end of Lake

Ontario, it was one of the front lines in the War of 1812 between Britain and the United States.

The USA had won its independence from Britain thirty years earlier, but Canada remained a loyal British colony. In 1812 the British were fighting the French Emperor Napoleon. The Americans remained friends with the French and were furious when Britain blocked American trade with France. In retaliation the United States declared war on Britain and invaded the nearest piece of British territory – Upper Canada. Very early on the morning of 13 October 1812 American troops crossed the Niagara River just downstream from the famous falls and headed north towards the small town of Queenston. If they could secure this foothold in Canada, the Americans would be well positioned to deal a strategic blow to the British Empire.

Laura Secord, a slim, brown-eyed woman, five feet four inches tall, with a friendly smile, lived above the family shop in Queenston with her husband James and their children. On the morning of the American invasion she woke up to the sound of gunfire and said she saw 'cannon balls . . . flying around in every direction'. Her husband, a member of the local militia, had gone off to fight with the British. So Secord hurried her children out of their clapboard house in the middle of town and took refuge in a neighbouring farm.

The British, under the command of Major-General Isaac Brock, were at Fort George, six miles north of Queenston. His troops included British soldiers, local men like James Secord along with Indians and a company of freed black American slaves. General Brock was an experienced commander who had distinguished himself by capturing Fort Detroit from the Americans two months earlier. As soon as he received the news of the American invasion, he ordered an advance party to confront the enemy.

The servant who strapped on his sword said, 'You are very early, sir.'

'Yes, but the Yankees are earlier,' replied Brock.

Brock led the way on his huge grey horse, Alfred. In his red coat, gold epaulettes and cocked hat, he was a perfect target and was soon spotted and killed by an American sharpshooter. The British were dismayed by the loss of one of their best military commanders. But hours later, even after the death of Brock, they managed to repel the Americans in what became known as the Battle of Queenston Heights. James Secord, who had helped to carry Brock's body away, was badly wounded. On learning that her husband lay bleeding on the battlefield, Laura rushed to find him. Most of the dead and wounded soldiers she saw were Americans wearing distinctive blue tunics but there were many British Redcoats too. She found her husband lying on the ground, a bullet in his knee and blood flowing from a shoulder wound. One of her granddaughters recalled Laura Secord's account of what happened next: 'Just as she reached the spot three American soldiers came up, and two of them raised their muskets to club him to death. My grandmother rushed in between them, telling them to kill her and spare her husband. One of them spoke very roughly and told her to get out of the way, and shoving her to one side, was about to accomplish his murderous intention.'

Fortunately an American officer intervened and arranged for the Secords to be safely escorted home. There another shock awaited them: American soldiers had looted their house. Laura Secord's anger turned into a deep hatred of the invaders.

The American attack on Queenston had been repulsed and freezing winter weather soon put the war on hold but in the spring of 1813 attacks on Canada took on a new ferocity. In April the Americans attacked the city of York (now Toronto) and burned down its parliament buildings and the British governor's house. A month later, the British were forced to abandon Fort George which controlled the entrance to the Niagara River after American troops crossed the river in large numbers. The Niagara Peninsula again became a war zone controlled partly by the Americans and partly by the British. The Americans occupied Queenston and this time it was to be Laura Secord, not her husband, who took them on. It was an action that would secure her a place in Canadian history.

On 21 June 1813 she learned of a planned attack on a British outpost at Beaver Dams, ten miles from Queenston. As well as fifty British soldiers, it contained crucial ammunition and supplies. Secord was at home when American soldiers knocked on the door and asked her to prepare a meal. She had no choice but to agree to their request and was shocked by their conversation. She later told her granddaughter that she 'listened outside the window where they were taking their supper and she overheard them say that they would surprise Colonel FitzGibbon at Beaver Dams'.

As soon as the Americans left her house, Secord told her husband what she had heard. He agreed someone had to warn Colonel FitzGibbon. Since James was still recovering from the wounds he'd received in the Battle of Queenston Heights, there was no way he could carry the message. Laura decided to make the perilous journey herself to 'save the British troops from capture and total destruction'. Her husband doubted that, with all the American soldiers around, 'any man could get through let alone a woman'.

Secord replied, 'You forget, James, that God will take care of me.'

At first light on 22 June one of her daughters glimpsed Secord leaving the house in a home-made brown cotton dress. She had a white muslin kerchief around her neck and a white cotton bonnet on her head. Her daughter said she looked like 'a housewife off on an errand'.

She went first to her half-brother's house nearby in the hope that he could carry the message to the British but he was sick in bed. So Secord felt she had no choice but to make the journey herself. She avoided the direct ten-mile route to Beaver Dams because there were American soldiers on that road. Instead she scrambled along a path, tripping over roots and wading through swamps. It had rained heavily that month so there was mud everywhere. Secord lost both her shoes. At one point she came to a river where the bridge had been washed away but she managed to crawl across on a fallen tree. The day got hotter and hotter. Badly bitten by mosquitoes and terrified at the prospect of running into wildcats or rattlesnakes, Secord pressed on doggedly. Her circuitous route meant she avoided American soldiers but, as darkness set in, she found herself facing people she, like many other colonists, dreaded – Indians.

It was common at that time to hear stories about Indians kidnapping white people. John Whitmore, who lived near the Secords, had been seized by Indians in New Jersey. They attacked his parents' farm, killed his father, mother and older brother and then forced four-year-old Whitmore and his six siblings to go with them. He lived with his kidnappers for four years until an army captain from Niagara rescued and then adopted him. When a family in Pennsylvania was carried off by Indians, relations of James Secord had paid for their release. And there was another reason for Laura Secord to fear Indians: some fought for the British, some for the Americans. Secord had no idea which side she'd stumbled upon. There was a chance that they would take her prisoner. She later described the scene: 'I found all the Indians encamped. By moonlight the scene was terrifying . . . They all rose and with some yells said "Woman" which made me tremble. I cannot express the awful feeling it gave me but I did not lose my presence of mind. I was determined to persevere.'

Fortunately these Indians supported the British. They didn't speak English but somehow Secord made their chief understand that she had an important message for Captain FitzGibbon who was in charge of the Beaver Dams outpost. The chief escorted her to FitzGibbon who could hardly believe his eyes. Here was 'a slight, delicate woman' in a torn dress, clearly exhausted, who had walked twenty miles to tell him that the Americans planned to attack within the next two days.

Along with the fifty men in his unit, Fitzgibbon had under his command 400 Indians who'd just arrived to join him. They were ready and waiting when 462 American soldiers crept furtively up on the British. The Americans had counted on a surprise attack but because of Laura's tip-off they found themselves in the middle of an ambush. As they walked through a beech wood three miles from their destination, the Americans were attacked from behind trees on each side of the road, then from the front and rear. In order to make them think they were surrounded, canny FitzGibbon got buglers to move around the woods. By blowing their horns from all directions they tricked the Americans into believing they were heavily outnumbered. Under cover of a white flag FitzGibbon then offered to negotiate. He told the Americans that he was about to be joined by a force of particularly bloodthirsty Indians whom he feared he'd be unable to control. He predicted a massacre. The Americans promptly surrendered.

Beaver Dams was not the most decisive battle of the War of 1812 but it had a crucial impact. It massively boosted British morale and damaged that of the United States. Within two weeks, the American commander responsible for the whole area from Niagara to the Atlantic coast was relieved of his command. The rout at Beaver Dams rekindled American fears of the Indians, and US patrols dared not venture more than a mile away from Fort George. By December 1813 the Americans had abandoned the fort and headed back across the Niagara River into the United States. By the end of the year British troops had crossed the river, captured Fort Niagara and burned the American cities of Lewiston and Buffalo. The war would continue for more than a year but the Niagara Peninsula remained firmly in British hands.

After her epic journey to Beaver Dams Laura Secord returned home to her husband and children. Over the years the Secords wrote several petitions asking to be recompensed for their war efforts. All were turned down. The memory of her gallant act might have been lost had it not been for the unlikely help of Prince Edward, the eldest son of Queen Victoria. Nearly fifty years after Secord carried her message across American lines, nineteen-year-old Prince Edward was touring British North America. He visited Niagara Falls in September 1860 and watched the French artiste Blondin walk across the rapids on a tightrope, carrying a man on his back one way and then back across the falls on stilts.

During his stay the Prince was given testimonials written by people involved in the War of 1812. Eighty-five-year-old Laura Secord's was the only account by a woman. The royal visitor was captivated by the story of how she had carried the message that saved the British army. Secord was at last to get her reward: when he returned to England, Prince Edward sent her £100. The generous gift made headlines in Canadian newspapers and people belatedly realised they had a star in their midst. The woman who had been known as James Secord's widow became Laura Secord the legend.

Over the years monuments were erected to celebrate her and schools and streets named after her. In 1913 an enterprising businessman established the Laura Secord chocolate shops which still exist all over Ontario today. Their original white boxes bore the image of Laura Secord as an elderly woman. That picture has since been replaced by a pretty young girl but one thing has not changed: Canada's Niagara heroine is still remembered with pride and affection.

6

Harry Percy

Percy's Purse, Waterloo, 1815

FOR THE PAST two centuries there has been speculation about who won the race to reveal that the Duke of Wellington had defeated the French Emperor Napoleon on 18 June 1815. The battle fought that day at Waterloo decisively ended decades of war that had ravaged Europe since the French Revolution in 1789. So news of the victory led to rapturous rejoicing in Britain and made a hero of the messenger. One popular tale is that the enterprising Nathan Rothschild or a spy hired by him learned the outcome first and made the banker an even richer man. There is no firm proof of this story or of the claim that Rothschild made a killing on the London Stock Exchange in the days following the battle. What is beyond doubt is that the official Waterloo dispatch, a detailed account written by Wellington, was carried to London by the Honourable Harry Percy, one of the duke's aides-de-camp. What's also true is that Percy carried the dispatch in a most unusual container: a purple velvet purse given to him by a secret admirer at the Duchess of Richmond's ball. It was at that glittering event on 15 June, three days before Waterloo, that Percy's extraordinary story begins.

The duchess was determined to make her ball the highlight of the Brussels social season. Little could she guess that her soirée would be immortalised by novelist William Makepeace Thackeray. 'All Brussels had been in a state of excitement about it,' wrote Thackeray in *Vanity Fair*. The ball was also celebrated in verse. Lord Byron's famous poem 'The Eve of Waterloo' begins: 'There was a sound of revelry by night' punctuated by 'the cannon's opening roar'. The timing of the ball was unfortunate to say the least. That day, 15 June, turned out to be the very day that Napoleon chose to invade Belgium. Defeated and forced to abdicate in 1814, he had been exiled to the island of Elba. But a year later he had escaped from exile, returned to Paris and now headed an army of 120,000 men intent on destroying the armies of the Duke of Wellington and his allies before they could attack him. For more than a decade and a half this formidable Frenchman had fought to dominate Europe: this was to be the decisive confrontation.

Fearing the worst, the duchess asked Wellington whether she should call off the event. 'Duchess,' he replied, 'you may give your ball with the greatest safety, without fear of interruption.' He and most of Brussels knew Napoleon's army was on the march, but Wellington wanted to be clear exactly where it was heading before he reacted. Napoleon was an old hand at strategic deception. Besides, as commander-in-chief of allied forces in Belgium, Wellington was determined to appear imperturbable and persuade the citizens of Brussels that there was no need to panic. The last thing he wanted was a mass exodus from the city while he was trying to deploy his soldiers to fight Napoleon.

So the guests, the men in all the glittering splendour of their full dress uniforms and the women twinkling with jewels in elaborate ball gowns, arrived at the Richmonds' house as planned on 15 June. They included a prince, sixteen counts and countesses, twenty-two colonels, Wellington and many of his officers – among them the dashingly handsome Harry Percy.

Twenty-nine years old with dark curly hair, the agreeable Percy had served in Wellington's army in Spain and Portugal as the British pushed Napoleon's forces out of the peninsula. He was grandson of the Duke of Northumberland, and son of an earl, but as the fifth son of the family he stood no chance of inheriting the title. Like so many other aristocrats of that time, Percy's father bought him a commission in the British army. Young Harry ended up in the 7th Regiment of Foot and he soon won the respect of the inspirational General Sir John Moore, who was to win the Battle of Corunna in 1809. He became Moore's ADC and was deeply attached to him. After the general was killed by a cannon ball at the moment of victory, Percy was a pall-bearer at his funeral. He carried a lock of Moore's hair for the rest of his life.

He stayed in the peninsula and impressed Wellington, the new army commander, who, like Moore, spotted his qualities and appointed him aide-de-camp. When Percy was taken prisoner by the French in 1810, Wellington took the unusual step of writing to the commander of the French force, Marshal Massena: 'Captain Percy, whose fate concerns me greatly, was wounded and captured or killed yesterday near Clerorico and I would be much obliged if you would allow news of him to be given to my aide de camp.' The French responded by allowing Percy to live under very relaxed house arrest with his parents at the home in central France they'd bought in peacetime. They were living there when war with France resumed in 1803 and were not permitted to return to England.

In the course of a fairly carefree three and a half years of house arrest Percy managed to produce two sons with his mistress Madame Durand. Once peace was restored in 1814 and Napoleon was banished to Elba, Percy, a fluent French speaker, was asked to join Wellington's staff in Paris where the duke had been sent as British ambassador. When Napoleon escaped from exile in early 1815 and Wellington was sent to command the opposing allied army in Brussels, he asked Percy to accompany him.

So Percy now found himself at the Duchess of Richmond's ball as Brussels waited for news of Napoleon's approach. As guests danced the night away, reports confirmed that Napoleon's armies were headed straight for Brussels. 'He has humbugged me, by God; he has gained 24 hours march on me,' lamented Wellington, ordering his army south at full speed to meet the French. Before Percy rushed off to prepare for battle he was handed a velvet purse, perhaps as a good luck charm, by an unknown dancing partner. He stuffed it into a pocket of the red dress coat he'd put on for the ball.

The next day Percy was by Wellington's side in the thick of the first clash with Napoleon at the crossroads of Quatre Bras and then on 18 June at Waterloo. His horse was killed but he escaped injury in the nine hours of fighting that ended with the repulse of the French Imperial Guard and the decisive, final defeat of the French emperor. When the battle was over, Percy witnessed the famous encounter immortalised in the painting by Daniel Maclise, *The Meeting of Wellington and Blücher*. Major Percy is portrayed just behind Wellington as he shakes the Prussian field marshal's hand in front of La Belle Alliance Inn, Napoleon's headquarters during the battle. He may well have overheard Wellington thanking the Prussian general for turning up at Waterloo in the nick of time, after Blücher greeted him with the words, '*Mein lieber kamarad, quelle affaire* . . . My dear friend, what a business.'

After meeting Blücher, Wellington and Percy rode to the duke's lodgings in the village of Waterloo. There the duke grabbed a few hours' sleep before sitting down in the small hours to write a sixteen-page report on the fighting – his famous Waterloo dispatch – to Lord Bathurst, British Secretary of State for War and the Colonies. With remarkable understatement he described the battle, praising individual soldiers who'd stood out and listing some of the many dead. Two of his more senior aides had been killed or injured and Percy was delighted when Wellington chose him to carry the dispatch along with 'two eagles taken by the troops in this action, which Major Percy will have the honour of laying at the feet of His Royal Highness'. The eagles were standards or flags captured from French regiments.

When Percy set off on Monday, 19 June, the day after the Battle of Waterloo, he was still wearing the red dress uniform he'd put on for the Duchess of Richmond's ball. It was now covered in the blood of a fellow officer, killed beside him at Waterloo. The purple velvet purse he'd been given at the ball was still in his pocket and he carefully placed the dispatch inside it. He travelled in a carriage drawn by two horses with the captured eagles poking out through the windows. Progress was frustratingly slow. Heavy rain had turned roads into muddy obstacle courses. They were also packed with soldiers and fugitives. Percy's first stop was Ghent, thirty miles from Brussels, where he told French King Louis XVIII, who'd taken refuge there, that his enemy Napoleon had been defeated. From Ghent he went to the Belgian port of Ostend, taking twenty-four hours to travel just seventy-eight miles. His triumphant arrival was witnessed by a British soldier, Benson Earle Hill. Rumours had been circulating that Napoleon had won the battle at Waterloo, so cheering crowds chased after the carriage when they caught sight of the captured French flags. 'A loud huzza was heard at some distance,' Hill wrote.

> A cabriolet drove up to the door, in which Major Percy was seated displaying to the hundreds who had followed him the eagles of the 45th and 105th regiments taken from the foe on the glorious plains of Waterloo . . . Major Percy alighted and, in a few words, related to the admiral the leading features of that fight upon which the destinies of Europe depended.

At 2 p.m. on Tuesday, 20 June, Wellington's messenger boarded a British warship HMS *Peruvian*. In the ship's log Captain James Kearney White reports leaving Ostend at 4 p.m. 'with the Hon. Major Percy with dispatches for the War Office from the Duke of Wellington'. Given fair winds, the seventy-mile crossing from Ostend to Deal on the coast of southern England would have taken as little as eight hours but the winds were anything but fair. They were scarcely blowing. Captain White's log says it all: 'winds East, light breezes' when they set sail and 'Light breezes, clear' at three in the morning. It was 'Nearly calm' at 8 a.m. By 11.30 a.m., after spending twenty anxious hours pacing the decks, Percy complained to Captain White that the voyage was taking far too long. 'Light airs, made all possible sail' reads the ship's log. But setting extra sails did no good. The ship was becalmed. Captain White, clearly a man of action, decided to take matters into his own hands – literally. He ordered that a boat be lowered overboard. 'Out gig', the log records, 'Captain and the Hon. Major Percy left the ship with the dispatches.' Percy had rowed as a schoolboy at Eton. He, Captain White and four strong seamen grabbed the oars and headed to England. They made it in surprisingly good time as the *Kentish*

Gazette newspaper reported: '4 hours after, [they] landed safely at Broadstairs; these gallant fellows having rowed 38 miles.'

Percy stepped ashore around 3.30 p.m. on Wednesday, 21 June. He and his fellow rowers walked through the village waving the eagles above their heads. The crowds who'd gathered round them were ecstatic. Here at last was the news a nation had been waiting for: Wellington's army was victorious. Percy stopped at an inn, later named Eagle House and hired a post-chaise and four horses to carry him as swiftly as possible to London, seventy-five miles away.

The Duke of Wellington had asked Percy to do three things when he reached the British capital: deliver his dispatch to Lord Bathurst; present the captured eagles to the Prince Regent, eldest son of King George III; and give a copy of the dispatch to the Duke of York, the king's second son and chief of the army.

It was nearly midnight when Percy's carriage pulled into Downing Street where he hoped to find Lord Bathurst. Minutes later Charles Arbuthnot, Secretary to the Treasury, who lived in Downing Street, was woken by shouts of 'Wellington is safe'. He looked out of his window and saw a crowd gathering around a carriage. He knew that Bathurst was expecting news from Wellington so rushed out and volunteered to take Percy to the War Minister who was having dinner with the rest of the Cabinet at a house in Grosvenor Square two miles away. As news of Wellington's victory spread, the noisy crowd following Percy's carriage grew ever larger.

The Cabinet was dining at the home of Lord Harrowby, Lord President of the Council. It was, by all accounts, a gloomy affair. Although Napoleon had been expected to attack three days earlier, no one had any idea what had happened. The Chancellor of the Exchequer, Nicholas Vansittart, described how:

> They dined, they sat. No dispatch came. At length, when the night was far advanced, they broke up. Yet, delayed by a lingering hope that the expected messenger might appear, they stood awhile in a knot conversing on the pavement when suddenly was heard a faint and distant shout. It was a shout of Victory! Hurrah! Escorted by a running and vociferous multitude, the Major drove up. He was taken into the house and the dispatch was opened.

When Harry Percy took the dispatch out of his purple velvet purse and handed it over to be read by the anxious Cabinet ministers, there was some confusion. Wellington didn't indicate who had won the battle until over halfway through. As Vansittart recounted, 'They could not exactly gather from a first reading of the dispatch on what scale the allied armies had been triumphant or how far the success was final and complete. They turned for information to Major Percy but the gallant officer was dead beat – much more inclined to go off into a doze than to answer questions.'

While Wellington's words were being minutely analysed, Charles Arbuthnot, who had accompanied Percy to the house, couldn't resist going outside to address the crowd. 'In short, the French army is entirely destroyed,' he shouted.

From inside came the voice of Lord Harrowby: 'I beg your pardon Mr Arbuthnot – but not exactly – I think you are going a little too far.'

It wasn't long before the true scale of Wellington's victory became clear, and the Prime Minister, Lord Liverpool, told the exhausted Percy, 'You must come with me immediately to the Regent.' The prince was attending a dinner followed by a ball at the home of the celebrated society hostess Dorothy Boehm. Her husband had made a fortune in the East India Company and the Boehms were hosting the evening at their elegant house in St James's Square in the heart of London (now the East India Club). For Mrs Boehm it was a dream come true that the Prince Regent and his brother, the Duke of York, had accepted her invitation and, as one newspaper reported, 'Mr Boehm had spared no cost to render it the most brilliant party of the season.'

But now this young major with a message from Waterloo was to turn her party so full of promise into a nightmare. In her own words: 'The first quadrille was in the act of forming and the Prince was walking up to the dais on which his seat was placed when I saw every one without the slightest sense of decorum rushing to the windows.' She described

> the vociferous shouts of an enormous mob who were running by the side of a post-chaise and four, out of whose windows were hanging three [there were actually only two] nasty French eagles. In

> a second the door of the carriage was flung open and, without waiting for the steps to be let down, out sprang Harry Percy – such a dusty figure – with a flag in each hand, pushing aside everyone who happened to be in his way, darting up the stairs, into the ballroom, stepping hastily up to the Regent, dropping on one knee, laying the flags at his feet and pronouncing the words 'Victory, Sir. Victory'.

Making little effort to shake off the dust and grime of the journey and fighting fatigue with a final burst of energy, Percy strode into the ballroom. He was followed by Lord Bathurst, the Prime Minister Lord Liverpool and most of the rest of the Cabinet. They carried Wellington's dispatch and the copy for the Duke of York and quickly disappeared into another room to brief the royal guests. The Prince Regent wept as Lord Liverpool read out the list of dead. With tears streaming down his face he lamented, 'It is a glorious victory and we must rejoice in it . . . but the loss of life is fearful and I have lost many friends.'

The prince and his brother promptly abandoned the ball followed by everyone else, leaving their hostess, Mrs Boehm, bereft and desolate. She remembered: 'Such a scene of excitement, anxiety and confusion never was witnessed before or since . . . The splendid supper which had been provided for our guests stood in the dining room untouched. Ladies of the highest rank, who had not ordered their carriages till four o'clock a.m. rushed away like maniacs, in their muslins and satin shoes.'

Within twenty minutes Mr and Mrs Boehm found themselves alone in their ballroom. To add insult to injury the noise of rejoicing crowds outside poured through their open windows. According to the *Courier*: 'They sang *God Save the King* full chorus and filled the place with shouts and huzzahs. A more animated scene cannot be conceived. The flags and the two eagles were displayed in front of Mr Boehm's house, and loudly cheered by the people, who pressed forward, anxious to touch them with their hands . . . All was joy and festivity to a late hour.'

For the rest of her life, Mrs Boehm blamed Harry Percy for ruining her ball.

> All our trouble, anxiety, and the expense were utterly thrown away in consequence of . . . the unseasonable declaration of the Waterloo victory! Of course, one was very glad to think one had beaten those horrid French . . . but I always shall think it would have been far better if Harry Percy had waited quietly until the morning instead of bursting in upon us, as he did, in such indecent haste.

His job done, Harry Percy went to his father's London home utterly worn out by his three-day journey. He took off his blood-stained coat and, according to family legend, as he unwound the scarlet sash from around his waist, bits of brain belonging to the officer who died beside him at Waterloo dropped at his feet.

Carrying Wellington's dispatch from Waterloo turned out to be the high point of Harry Percy's career. He suffered ill health and left the army in 1821 and became a Member of Parliament. There is no record of him ever speaking in the House of Commons. He only voted once. He died in 1825 when he was only thirty-nine. He never married his French mistress Madame Durand although the two sons they had together moved to London. One of them, Henry Durand, went on to become a major-general in the British army.

SPECIAL FORCES

THERE'S AN UNDENIABLE mystique about Special Forces that makes them a permanent source of fascination. Their popular image as undercover hitmen and women with a licence to kill, and the secrecy that surrounds their identities and assignments only enhances their reputation as cloak and dagger operatives. But most Special Forces are in fact fighting troops trained to the highest levels of combat skill and endurance. Their task is to inflict damage on an enemy in all sorts of unconventional ways, often far behind the lines.

Specialist troops such as the United States Rangers have had a role since the seventeenth century but it was the Second World War that prompted several nations to set up their own Special Forces. Winston Churchill, Britain's wartime Prime Minister, decided he needed what he called 'specially trained troops of the hunter class, who can develop a reign of terror'. This led to the creation of the Commandos, one of whom, Major-General Corran Purdon, told us how he took part in what's been called the greatest raid of all – the successful but costly attack on the port of St Nazaire in Nazi-occupied France. Of the 611 men who set off, only a third got safely home. Churchill's measures were mirrored in other countries around the world. The German army's Jäger Battalion under Otto Skorzeny and the Japanese 'Raiding Groups' caused the Allies severe anxiety in the Second World War. The Soviets invented their Spetsnaz units to worry NATO in the Cold War. US Navy Seals, conceived in the early 1940s, staged the raid that led to the killing of Osama bin Laden in 2011.

Perhaps the most famous of all Britain's elite military units is the SAS, the Special Air Service. It too began its life in the Second World War. It was dreamed up by a Commando called David Stirling who was convinced that he could lead a small group of men to carry out attacks behind enemy lines in the desert campaigns in North Africa. General Montgomery, Britain's successful desert commander-in-chief, described Stirling as 'mad, quite mad'. But he added, 'in war there is often a place for mad people.' One of these 'madmen', Mike Sadler, described to us his extraordinary memories of what it was like to be a member of the SAS. With their motto 'Who Dares Wins', SAS fighters staged hundreds of key raids in the Second World War and have gone on to operate in Malaya, Aden, Northern Ireland, Afghanistan, Iraq and even in London where they rescued hostages from the Iranian Embassy in 1980. The aura surrounding the SAS is deepened by the fact that no British government ever comments on its activities.

Churchill created another special force even more secretive than the SAS. It was called the Special Operations Executive, the SOE, and it was ordered by Churchill to 'set Europe ablaze'. SOE agents, posing as civilians, infiltrated countries occupied by Nazi Germany and in the Far East. They blew up bridges and power stations and assassinated key figures such as Reinhard Heydrich, second-in-command to Heinrich Himmler, Hitler's right-hand man. Many SOE agents were women which was highly unusual for the time. As well as recounting the stories of Mike Sadler and Corran Purdon, we have chosen to highlight one of those SOE female operatives, the astonishingly brave Krystyna Skarbek.

7

Krystyna Skarbek

Spy, Saboteur and Enchantress, 1939–45

IT IS HER captivating smile that draws us to Second World War spy Krystyna Skarbek. In photos taken on dangerous assignments, she radiates beauty, vitality and happiness. Behind that smile was a fearless woman of ruthless cunning prepared to fight to the death to defeat Hitler's Nazis. She was Winston Churchill's favourite spy. Colleagues described her as 'brave as a lion' but with 'a positive nostalgia for danger', and 'an almost pathological tendency to take risks'. She refused to give up, no matter how great the task. She skied over mountains into German-occupied Poland to gather intelligence on the Nazi occupation; she set up escape routes for Allied servicemen from behind enemy lines; she blew up bridges; she risked her life rushing to the rescue of jailed resistance fighters hours before their execution. In one of many lucky escapes she charmed a snarling German guard dog by lying beside it and scratching its ears. She was irresistible to men, including author Ian Fleming and was said to have been the inspiration for his first Bond girl in *Casino Royale*. Fleming was just one of her many lovers. Her reckless use and abuse of the men in her life led to her premature and tragic end.

Skarbek was Britain's first female agent, the longest serving and arguably the most effective in helping undermine Hitler's grip on occupied Europe. Born in Warsaw in 1908, she enjoyed a privileged upbringing. She was headstrong, with a love of adventure, and attracting men. By the age of thirty she'd been married twice.

After Germany invaded her beloved Poland in 1939, Skarbek was so appalled that she made her way to London and volunteered to spy for Britain. Foreign Office records reveal that 'a flaming Polish patriot, expert skier and great adventuress' offered to go to Hungary (still neutral in the early part of the Second World War) and then ski into Poland carrying anti-Nazi propaganda leaflets. She would send back intelligence reports and help Allied prisoners of war escape. 'She is absolutely fearless,' wrote a British official. 'We have a prize.' Skarbek became an operative in Section D – short for 'Destruction' – which later became the Special Operations Executive, the famous SOE.

Skarbek was sent to the Hungarian capital Budapest in December 1939 posing as a French journalist. She didn't waste any time. As well as preparing propaganda leaflets, she set up clandestine radio broadcasts to Poland and stored explosives in her apartment. She also had a very active love life. When reports of her romantic adventures reached head office, Section D reported primly: 'Her attractiveness appears to be causing some difficulty in Budapest.' Her most unfortunate admirer was Polish journalist and fellow British spy Jozef Radziminski. When she rejected his overzealous advances, he tried to commit suicide by jumping off a Budapest bridge but landed on the frozen Danube River, breaking his collarbone. He then threatened to shoot his genitals off in front of Skarbek but, panicking at the last moment, ended up wounding his thigh.

Although still married to her second husband, George Gizycki, Skarbek fell in love with a one-legged Polish resistance fighter named Andrzej Kowerski. Their relationship was one of the most important of her life, both in bed and in the field. They were a brilliant team, credited with saving the lives of hundreds of Allied servicemen. It was Kowerski who persuaded a former Polish Olympic skier to guide Skarbek from Slovakia, over the 2,000-metre Tatra Mountains, and smuggle her across the border into Poland. This was a daunting task even for an experienced skier like Skarbek, who had once survived three and a half hours buried in an avalanche. She and her guide zigzagged up a mountainside through snow five metres deep. It was minus thirty degrees and a blizzard forced them to take refuge for the night in a mountain hut. Later as the storm got worse Skarbek thought she heard distant voices crying for help. She rushed to the door anxious to go to the rescue but her guide forced her back, saying, 'I'm sorry, Krystyna. We have important work to do and I cannot jeopardise our lives by going out into that raging snow.' The next day they found the bodies of a man and young girl lying frozen on the snow. They had been trying to escape from Poland. Thirty people died on the mountain that night.

It took three days to reach the Polish city of Zakopane, but her adventures had only just begun. On the train to Warsaw Skarbek noticed German guards checking passengers. A Gestapo officer stopped her and asked about the parcel she was carrying. It contained incriminating documents but with typical sangfroid she told him it was black-market tea for her sick mother. (Skarbek's mother Stephanie was Jewish, the daughter of a rich banker. She was arrested by Gestapo agents in Warsaw during the war and is thought to have died in a concentration camp.) The officer offered to safeguard the parcel by putting it in his suitcase. When the train arrived in Warsaw, he graciously handed it over, blissfully unaware that he was helping the enemy.

Skarbek worked with Polish resistance groups using an underground press to print propaganda. She sent reports back to London about 'the terrible disorder of the German administration', describing how open antagonism between the German military and the Gestapo led to contradictory orders being issued every day. Like so many other eyewitnesses, she confirmed that Jews were being mistreated. Once she nearly betrayed herself by stopping to watch in horror as a German officer pushed an elderly Jewish man into the road with the end of his machine gun. When the officer asked for her identity papers she forced herself to hand them over with a smile.

'In Warsaw alone over 100 Poles are shot every night,' she wrote.

'The terror is indescribable. Yet the spirit of the Poles is magnificent.' Her fury at what the Germans were doing to her country became Skarbek's driving force. In the words of one of her fellow spies she became 'an avenging angel'. Travelling across Poland by train and on foot, she sent the British reports about German troop and equipment movements. She found getting in and out of Poland increasingly dangerous as Hungary, where she remained based, and neighbouring Slovakia edged closer to Hitler and the Germans tightened up border crossings. On one occasion Skarbek and a Polish resistance fighter (who also became her lover) were shot at as they tried to cross into Slovakia. Fortunately the sun was shining into the border guard's eyes and he missed.

Skarbek was a genius at avoiding capture. Her beauty, charm and complete disregard for personal safety got her out of many potentially fatal encounters. On one occasion she and another spy were being marched to a Slovakian police station for questioning. She distracted their guards which gave her companion time to throw incriminating documents into a nearby river. The guards tried unsuccessfully to fish the papers out, then turned on their captives and searched their belongings in a fury. When one came across a glass necklace in Skarbek's knapsack, she spotted a chance to escape. Scratching at the guard's face, she screamed there were diamonds in the necklace and grabbed at it. It broke, and as the guards scrabbled in the dirt for what they thought were precious stones, Skarbek and her companion ran for their lives. They escaped in a hail of bullets but their knapsacks containing false identity papers were left behind. Skarbek's picture was posted offering a reward of £1,000 'dead or alive'.

Undaunted, she continued working with Andrzej Kowerski, finding couriers to move arms and money into Poland and carry out intelligence. They also helped smuggle people along what became known as 'the underground railway'. Skarbek shepherded groups of people from Poland and Slovakia to Hungary, and Kowerski arranged for them to be driven from Hungary to safer countries in southern Europe. According to British estimates, they helped 5,000 people escape. In a lucky break, Kowerski arranged to smuggle out relations of a disaffected Hungarian colonel and received an official car and a military pass as payment. That meant he and Skarbek could travel freely around Hungary collecting microfilm and other intelligence. They hid the top-secret material in her gloves or in his wooden leg. They engaged in sabotage operations too. Once Kowerski and a colleague placed limpet mines on a Danube River oil barge heading through Hungary to Austria. For once Skarbek was not in the thick of the action as she couldn't swim so waited in a getaway car holding her lover's indispensable wooden leg. Sir Owen O'Malley, the British ambassador in Budapest, was so impressed by Skarbek's daring espionage that he wrote: 'She was the bravest woman I ever knew . . . she could do anything with dynamite except eat it.'

Given how active they were, we are not surprised that Skarbek and Kowerski were put under surveillance. In the early hours of 24 January 1941 their Budapest apartment was raided by military police. Skarbek was asleep in bed, stark naked, but far from acting terrified, she appeared to welcome the intruders. She gaily offered them a cup of tea. Kowerski later said, 'She looked as merry as if she were going to a cocktail party.' Before she stepped out of bed, she bashfully asked the policemen to avert their eyes so that she could get dressed. That gave her time to nip into the loo and flush down incriminating material. The two of them were taken for questioning by the Gestapo and Skarbek once again showed – British records say with typical understatement – 'great presence of mind'. She had recently recovered from flu and still had a hacking cough. She bit her tongue and pretended to cough up blood. The Germans feared she had highly infectious tuberculosis and called a prison doctor to examine her. He ordered an X-ray which showed ominous shadows on her lungs and promptly diagnosed tuberculosis. Terrified they'd be infected, her interrogators sent her and Kowerski home. Skarbek knew full well that the scars on her lungs had been caused fifteen years earlier by fumes at a Fiat car showroom where she had worked.

SOE now judged the pair were being too closely watched for comfort, and British ambassador O'Malley arranged for them to escape from Hungary, Skarbek in the boot of his official car. Before they left, the ambassador provided false passports and visas. Skarbek chose the name Christine Granville – a name she would use from then on. She also came up with a new birthdate, 1915, which shaved seven years off her real age. She stuck with that date for the rest of her life. The couple made it safely to Belgrade and it was there that Skarbek achieved one of her greatest coups. She was handed several rolls of microfilm by a Polish courier which she passed on to British intelligence. They contained pictures of German troops and equipment moving towards the Russian border. This was the first evidence that Hitler was preparing to attack his Russian ally and it quickly ended up on the desk of Winston Churchill, Britain's wartime leader. He later wrote: 'that Germany should at this stage, before clearing the Balkan scene, open another major war with Russia seemed too good to be true.' As documentation of Germany's invasion plans mounted, Churchill sent Stalin a telegram in April 1941, warning that Germany was about to turn on Russia. But convinced that this was an Allied plot aimed at ending his pact with Germany, Stalin refused to believe Britain's Prime Minister. He could not have been more mistaken. Two months after Churchill's warning, Hitler's armies rolled across the Russian border.

In spite of providing the first proof of the planned German invasion of Russia, Skarbek fell into disfavour. A Polish intelligence report claimed that her 'indiscretions' had led to the death of at least one Polish agent. She was also wrongly suspected of being a German double agent. Keen to remain on good terms with the Poles, the British security service distanced itself from Skarbek who was now living in Cairo. The hyperactive spy was suddenly without work. 'I am so fed up with everything and everyone that I am becoming a desert island,' she wrote to a friend. She complained she had been 'put on ice'. One contemporary noted that although her spying stopped for a time, she continued to have a very active love life. Men in Cairo 'found her terribly attractive but women would be rather fed up about the fact that she had such an effect on men'. War hero and writer Paddy Leigh Fermor remembered 'her beautiful looks' and her 'patriotism . . . backed by a love of adventure'.

It wasn't long before the British re-engaged Skarbek. Starting in the autumn of 1942, she trained as a wireless operator, and a small arms expert. She finally got the chance to try out her new skills in July 1944 when she was dropped by parachute into Nazi-occupied France. She carried a Commando knife, a revolver, a compass hidden in a hairclip, a map printed on a silk square and a deadly cyanide pill to swallow if captured. Her main mission was to work as a courier for British spy Francis Cammaerts who was in charge of resistance groups in south-eastern France. They soon became lovers. Skarbek organised the retrieval of weapons and explosives dropped from Allied planes and hid them until they could be used to blow up bridges and railway lines. She carried messages through German lines from Cammaerts to the Maquis – local French resistance fighters. Once she was stopped by a Nazi patrol while carrying her silk map of the area. She calmly shook the map out and used it as a scarf to tie back her hair. Another time two German soldiers blocked her path as she led an Italian resistance fighter to join the Maquis. Skarbek defiantly grabbed two grenades from her pack and, waving one in each hand, threatened to blow herself and the Germans up. Terrified, the two men let her and her companion go. During Operation Toplink – set up by the Allies at the time of the D-Day landings in June 1944 – Skarbek spent weeks climbing and crossing mountains to coordinate the activities of partisans in the French and Italian Alps. She sent back reports on partisan strikes and enemy troop movements. Toplink officer Paddy O'Reagan called her 'the gayest and most alive person I have ever met . . . With women like that around, there was nothing to do but bottle up one's incompetence and go through with it.'

Somehow, in the midst of all her activities, Skarbek found time to encourage the surrender of Polish soldiers who'd been forced to fight in the German army. It was a task that became easier after the successful D-Day landings in Normandy – landings that made a German defeat look more and more likely. In early August 1944 she learned that an Alpine fort at Col de Larch was full of Poles. The fort dominated a 2,000-metre pass used by the Germans to move troops between Italy and France. Skarbek spent two days climbing up steep animal tracks through dense forest with a megaphone slung over her shoulder. Wearing a scarf in Polish colours of red and white, she shouted commands to sixty-three fellow countrymen inside the fort. It's not clear why she was not shot by German guards (perhaps they thought she was mad) but for some reason she was able to tell the Poles that once it became clear that Allied armies were on their way, they should desert and join French resistance fighters. She even handed out typed instructions about how to surrender and which weapons to bring with them. The Allies landed in southern France on 15 August 1944 and the Polish troops deserted the fort four days later. They followed Skarbek's written instructions to a T, stealing the garrisons' arms. All sixty-three Polish soldiers joined the French Resistance.

Getting Polish soldiers to defect seemed like child's play compared to Skarbek's next heroic act. While she was busy in Col de Larch, the Gestapo had arrested Francis Cammaerts and two other resistance fighters. By the time she got back to her base at Seyne in south-east France, they had been condemned to death. She tried to persuade local resistance fighters to launch a rescue attempt but they said it was too risky. Always resourceful, Skarbek bicycled twenty-five miles to the garrison at Digne and arranged a meeting with Max Waem, who worked for the Gestapo. Claiming to be Cammaerts's wife, she announced she was a British spy, and produced some crystals from her wireless set. She told Waem that she was in constant touch with Allied forces which were on the point of landing. For good measure, she said she was the niece of Field Marshal Montgomery who commanded Allied forces in the D-Day operation and warned that 'the reprisals would be swift and terrible' unless Cammaerts and his comrades were released. Skarbek later said it was 'a stab in the dark' but Waem fell for it. Demanding a huge ransom, he promised to make the necessary arrangements for the condemned men's release. The cash was parachuted in on 16 August and a day later Cammaerts and his comrades were marched out of their cells by an armed Gestapo officer. Cammaerts, assuming they were about to be shot, thought, 'This is it.' But instead of being lined up in front of a firing squad, the trio was bundled into a car. Skarbek was in the front passenger seat. As they drove off, she smiled and said, 'It's worked.'

Cammaerts sent the SOE a report of how Skarbek saved his life. On it colleagues wrote comments such as: 'I am going to keep on Christine's side in the future' and 'So am I. She frightens me to death.'

For her ingenious and daring rescue, Skarbek was awarded the George Medal for outstanding bravery in action in France. The citation stated: 'The nerve, coolness and devotion to duty and high courage of this lady which inspired a successful conclusion to this astonishing coup de main must certainly be considered as one of the most remarkable personal exploits of the war.'

Sadly the rest of Skarbek's war proved frustrating and demoralising. There were plans to drop her into still occupied Poland but they were cancelled. 'The end of the war intervened,' it says in her file. Francis Cammaerts and other British agents were sent behind enemy lines in Germany. But not Skarbek: she offered to go there too to 'get people out from camps and prisons . . . before they get shot'. But the SOE did not permit female spies to travel to Germany.

Her treatment by the British authorities after the war was nothing short of shameful. She was left without a job, money or a homeland. She moved back to Cairo and friends in the SOE tried to get her work. 'She cannot type, has no experience whatever of office work and is altogether not a very easy person to employ,' was one official observation. British historian Alistair Horne met Skarbek in Cairo in 1945 when he worked for MI6. He told us, 'She could have been a valuable operative,' and was surprised that the British government didn't keep her on as a spy. Skarbek told him she was so broke that she had to sell her flat in order to survive. He remembered how witty she was and how strikingly beautiful, 'turning heads at the Gezira Club' where she sunbathed surrounded by admirers. He was struck by her long legs covered in scars from her wartime exploits and noticed 'like so many agents she never talked about what she did in the war.'

Skarbek's attempts to get British citizenship were also dealt with in a callously offhand way. Memos called her application a 'headache', and she was referred to as 'this girl'. She was told under British law married women could only apply for citizenship through their husbands. Skarbek had not seen her husband for years and he now lived in Canada, but in order to become a British citizen she had to divorce him. Even then, things were not ideal. When she finally got her passport it said she was a 'British subject by naturalisation'. She complained that made her 'feel like a convict' and created such a fuss that the Home Office changed the wording. Thanks to Skarbek 'British subject by naturalisation' was replaced by 'Citizen of Great Britain and the Colonies'.

Like many of Britain's demobbed service people who faced the test of peacetime, the spy whose wartime heroism had earned her an array of awards including the OBE, the George Medal and the Croix de Guerre ended up in a series of dead-end jobs. She worked as a hat-check girl, sold dresses at Harrods, and waited on tables in a Polish café in London.

Her end came violently. The woman who had narrowly escaped death on countless occasions in wartime was not so lucky in a post-war world. In the spring of 1951, Skarbek, or Christine Granville as she now called herself, signed on as a steward aboard a luxury ocean liner, the *Ruahine*. Paddy Leigh Fermor wrote poignantly of his dejected friend: 'She sailed away from her rather thoughtless, adopted country out of restlessness, independence and need, and at a loss suddenly with no one to rescue.' Skarbek's life at sea brought drama and tragedy. She had an affair with a fellow steward on the ship, Dennis Muldowney. A year later when she told him their affair was over, he stabbed her to death in the dingy hall of a London hotel. It was a tawdry end to a glorious life.

8

Mike Sadler

SAS Driver and Navigator, 1941–5

MIKE SADLER BLINKED his eyes open in the bright Sahara sunlight. 'I was looking straight down the barrel of a gun held by a man in the Afrika Korps.' He was twenty-three. After three years rampaging with the Special Air Service way behind Rommel's lines in the western desert he was ready for anything, but this was a shocker. When the German turned away to arrest more of Sadler's comrades, he leapt out of his sleeping bag and made a run for it. He had no idea that he was in for a footslog across 100 miles of desert – with no food or water. The story of his escape, which he told us more than seventy years later in the comfort of his apartment in Cheltenham, left us astonished that he survived.

Mike Sadler joined the Second World War's most audacious desert adventurers in 1941. When war broke out in 1939 he was nineteen, a Gloucestershire lad scarcely out of school. Anxious to be in the action, he enlisted as an artilleryman in East Africa. He was later posted to Egypt, a vital British base in the Second World War, under serious military threat from Germany and its Italian ally. One day

Sadler happened to be chatting in a Cairo café with some young soldiers who told him they were in the newly formed Long Range Desert Group, the LRDG. Their task was to drive their light vehicles well behind enemy lines and gather intelligence. They needed an anti-tank gunner. Would Mike like to join them? 'It sounded exciting,' recalled Sadler. 'I was worried I might miss fighting if I didn't go: besides, travel and adventure were in the blood. My father went off to the Yukon at an early age.'

Within days Sadler was driving south-west deep into the desert to Kufra oasis 700 miles away in southern Libya. To the north, along the coast of the Mediterranean, Britain and the two Axis powers were fighting a bloody battle that was shifting in the enemy's favour. General Erwin Rommel, dynamic commander of Hitler's Afrika Korps, whose swift mastery of warfare in the sands of North Africa had earned him the nickname the 'Desert Fox', was determined to break through to the Suez Canal with his Italian allies and sweep the British out of Egypt. The LRDG had an important role in trying to thwart him.

As his patrol headed south across the sand to Kufra, Mike Sadler was fascinated by the work of the team's navigator. By day he used a compass, but the key moment came after dark when the column of trucks stopped for the night. Sadler watched the navigator unload his theodolite, place it on its tripod, and take a star sight that would establish their position. He was so obviously captivated that he was asked if he'd like to be a navigator too.

'Oh yes,' he told us. 'I was very glad to abandon my anti-tank gun.' Sadler recalled how he spent two weeks at the oasis learning how to fix his position no matter where he was in the vast and featureless expanse of the Sahara Desert. It was a very exacting science, but he found that, being young and keen, his brain was quick to grasp its principles. He had to identify four stars, preferably in different parts of the heavens, and use the theodolite to establish their precise relationship to each other. A look at his watch then allowed him to note the exact time of that configuration and consult the almanac that was always carried in the navigator's kit. And then he discovered to his immense satisfaction that he could plot his latitude and longitude on a map with an accuracy of less than half a mile. It wasn't long before Sadler had proved himself the best navigator in the team. 'I wasn't brilliant,' he told us, 'I was just lucky.'

Within weeks of learning his new trade Sadler found himself driving off across the desert with his LRDG teammates to meet up with an undercover patrol. Their operations behind enemy lines would soon make them legendary heroes. In December 1941 Sadler's team drove to an oasis called Jalo 200 miles south of the Libyan coastal strip where Rommel and the British were fighting a fierce set of battles. Into Jalo also drove a patrol of the newly formed unit led by two young officers. Relatively unknown, they would be household names by the end of the war. Their commander was David Stirling, a quietly spoken but supremely self-confident twenty-six-year-old, who a few months earlier boldly walked into the office of Britain's commander-in-chief in Egypt and persuaded him to arm and supply a small force of raiders who would carry death and destruction way behind Rommel's lines. They were to be called the Special Air Service, the SAS.

Another SAS officer, with whom Sadler immediately struck up a close partnership, was Paddy Mayne. He was tall and powerfully built, an Ireland rugby international. He had a fearsome reputation as a ruthless fighter who would stop at nothing. 'He really enjoyed firing guns at people,' Sadler told us. 'Most of the time he exercised an immense but quiet authority. "Watch your language," Mayne would say to troopers who used swear words. He'd been brought up by his straight-laced mother in Newtownards.' Sadler described him as 'a smouldering volcano'. Stirling said Mayne was a man of 'satanic ferocity'. He would fight like a tiger on operations but would sometimes lose control off duty. Sadler later witnessed Mayne ending a convivial lunch party by taking a grenade from his belt, pulling out the pin and slapping it down on the tabletop. Everyone except Sadler dived under the table. 'I knew he wouldn't blow us up . . . and in fact he'd taken out the detonator.'

The LRDG immediately teamed up with the SAS and were nicknamed the 'Desert Taxi Service'. But Sadler was nothing like a London cabby. He was the navigator and guide on the first of the SAS's successful raids which dealt massive destruction on the airfield of Wadi Tamet, a key German and Italian base on the coast road.

Their vehicles mounted with four machine guns, and stashed with demolition charges, Mills bombs, jerry cans of petrol and water, and rations for several days, they drove some 500 miles to their target. 'It was Mike's first patrol as navigator,' said one trooper. 'In spite of the rough going and darkness he made it dead on. A marvellous performance he was to repeat often.'

Sadler helped Paddy Mayne and his four-man raiding party load up with a whole arsenal of bombs, grenades and a machine gun each, then watched them join a handful of other SAS men and disappear into the darkness. It was two hours before quite suddenly the sky lit up with a series of explosions. Mayne and his team had closed in silently on a mess hut full of partying Italian and German officers. Mayne kicked the door open, drew his Colt .45 pistol and stood there flanked by two of his men with sub-machine guns. He said, 'Good evening', and the two sides stood staring at each other until one German moved. Mayne and his companions then loosed off a fusillade that left more than two dozen dead. Leaving part of his team to finish the job, Mayne and a demolition team went off to destroy a total of twenty-four aircraft. They placed their time-fused bombs on the wings of as many aircraft as they could – German Stuka dive-bombers, Messerschmitts and Italian CR 42 fighters. When they ran out of explosives they used their guns to disable the planes, and finally with one Messerschmitt still unattended to, Mayne climbed into the cockpit, tore out the instrument wires and smashed the panel.

All the SAS men returned safely and moved off with Sadler and the LRDG. By the time they had driven a few miles the sun was up and Sadler remembered two Italian biplane bombers roaring in to attack them. 'Fortunately we could see the bombs dropping on us and I had time to swerve off to the right or left.' But when the biplane pilots used their machine guns it was harder to escape, because the planes were so slow they could turn quickly and keep their targets in sight. But somehow Sadler got away.

For over a year Mike Sadler guided Mayne and Stirling to their targets. His night navigation never failed, although bad weather could obscure the stars and force him to rely on his compass and dead reckoning. After one long night drive towards an enemy airfield

Stirling became impatient and demanded of Sadler, 'Where is this bloody place?'

'Just ahead,' replied Sadler, keeping his fingers crossed.

And at that moment just a few hundred yards ahead of them the Germans switched on a huge set of floodlights, illuminating the runway for a plane to land. 'You can't believe how satisfying that was,' recalled Sadler.

After a series of clashes in which Rommel's Afrika Korps and the British Eighth Army seesawed back and forth in the desert, Rommel came close to breaking through at El Alamein in the summer of 1942. But his supply line was over-extended and vulnerable to Allied air attack and to raids from Stirling's SAS. General Bernard Montgomery's massive build-up and skilfully prepared offensive of October and November forced Rommel into a long-drawn-out final retreat. It wasn't until May 1943 that the Afrika Korps was finally crushed in northern Tunisia.

Throughout this time Sadler was busily navigating Stirling's men behind German lines. At the beginning of 1943 with Rommel withdrawing into Tunisia, Stirling planned one of his most audacious operations yet. He and his team – this time without Mayne – would thrust into Tunisia, hit the Germans whenever they saw an opportunity, and then link up with the Allied First Army advancing from the west. Stirling told Sadler he was appointing him a fully fledged officer in the SAS. 'Better get yourself some officer's pips,' he said. Sadler promptly went off and bought some in a shop in Cairo. 'It was all completely unofficial,' Sadler told us, 'and when I was next in Cairo a military secretary told me, "I hear you're masquerading as an officer." But they let me keep my commission.'

The Tunisian operation was to turn into a horrific ordeal for Sadler. It would also end Stirling's wartime career. In January 1943 after a long and exhausting drive from Egypt, Sadler navigated the SAS team into southern Tunisia. Once through German lines they were flanked by difficult terrain and decided to bluff their way up the main road. Their fleet of jeeps drove past the men of an entire German armoured division who were lying down enjoying a coffee break in the morning sun. The German tank crews glanced curiously at the scruffily dressed men bumping along in their dust-covered vehicles but did nothing. More worrying was a German reconnaissance aircraft that made a menacing pass over them as they headed north. Stirling and Sadler decided to peel off the main road and lose themselves in the desert. They scattered around an empty watercourse to pitch camp, camouflage their jeeps and get some well-earned rest. Sadler and two comrades lying near him were too fatigued to keep a watch. They were soon in a deep sleep.

It was lunchtime when some sixth sense made Sadler open his eyes. He found himself looking up at a German soldier with his finger on the trigger of his rifle. For a moment or two he saw himself either dead or in a Nazi prison camp. But then astonishingly the German moved away and ran off to help round up the main body of the British patrol further down the hill. Sadler didn't think twice. Together with two others he leapt up without pausing even to grab his gun or any food or water and raced up the hill. Miraculously the Germans were so busy arresting Stirling and the rest of his men that they didn't spot the runaways. The three men dived into a slight depression in the scrub. Sadler, lying as still as he could, scraped out a hole and buried a pocketful of top-secret telegrams he feared the Germans could decode if they got their hands on them. He could hear the Germans firing and shouting but then he heard something much closer: a munching sound. The three were soon surrounded by a very large and smelly herd of goats. Whether the Arab goatherd had spotted their predicament and moved his animals to shield them they would never know. But the Germans did not find them and half an hour later the goats had gone and the place was silent.

Sadler and his two companions had only a very small-scale map and a compass. They debated whether to go back the way they had driven into Tunisia, but that would mean almost certain capture. They decided to head west across empty desert. It was more than 100 miles to Tozeur in western Tunisia where they knew they'd meet Allied forces. They had no food or water and the trek would take them at least three days. They ran into a group of Arabs who, instead of offering them water, demanded their clothes and began stoning them when they refused to hand over their battledresses. One of Sadler's mates was hit on the head and they had to drag him away as best they could. They noticed none of the Arabs was wearing shoes so fled across a wide bed of very sharp stones. They managed to staunch the blood from the wounded man's head and were lucky enough to spot another Arab encampment. This time they got a friendlier reception when they explained they were escaping from the Germans. They left with some dates and a leaky goatskin of water. On they went, close to exhaustion, and finally, after four full days' walking, they happened upon a Free French patrol who found it hard to believe they had crossed Tunisia and were from Britain's Eighth Army. 'The French handed us over to the Americans,' said Sadler. And he added with typical understatement, 'The trip had on the whole been quite frightening.'

Mike Sadler did not see David Stirling again until after the war. The SAS commander was among those captured while Sadler escaped and spent the remaining two years of the war a prisoner in Colditz Castle. SAS leadership was transferred to Paddy Mayne, and with the Germans cleared out of Africa their operations soon centred on occupied France. After D-Day in June 1944 Mayne took Mike Sadler with him on several raids on the Continent. By the end of the war Sadler had won himself a Military Cross and Mayne no fewer than four Distinguished Service Orders.

At the time of writing Mike Sadler was ninety-five, and enjoying a long retirement after a second career working for the Foreign Office. When we asked him what his job there had been he told us he would rather not go into it. Perhaps he will let us write that story one day.

9

Corran Purdon

Commando in 'the greatest ever raid', St Nazaire, 1942

IN THE SUMMER of 1939 an eighteen-year-old boy packed up his collection of lead soldiers and went off to learn to be a real one. He enrolled as an officer cadet at Sandhurst Royal Military College just two months before war broke out with Hitler. Within two and a half years, still only twenty, he would embark on what has been called the most spectacular raid of the entire Second World War. We heard about it in detail when we visited Major-General Corran Purdon, ninety-four years old, at his home in Wiltshire. His large, rambling house with its peaceful garden was a striking contrast to the savage scenes he described seven decades earlier.

Purdon, born in Ireland into a military family, dreamed of going into action as soon as his father left him at Sandhurst's gate. He had no idea his opportunity would come so soon. When Hitler's army invaded Poland on 1 September 1939 Britain declared war on Germany. It wasn't long before Britain's war effort was demanding a vast intake of new young commanders and Purdon's adventurous spirit made him a natural recruit for the Commandos, an elite fighting force. He was immediately immersed in a vigorous routine of fitness and weapon training. But two frustrating years passed during which nearly every other unit seemed to be fighting except his. He was sent on one raid to German-occupied Norway which was abandoned when his Commando carrier ship broke down. 'We were completely pissed off,' he told us. 'Why did we join the Commandos if all that we do is get sent nowhere?'

In February 1942 Purdon was sent off to a demolition course in Scotland, where he was taught about dry docks where ships were brought in to be outfitted or repaired. He learned about the docks' huge lock gates called caissons. They were made of concrete and opened and shut by massive engines in winding houses at each end of the lock. He also learned how to carry, place and detonate plastic explosive. 'All of us became adept at the various means of ignition,' recalled Purdon. He remembered wondering what all this training would lead to. He had no idea that he would be part of Operation Chariot, a daring offensive being plotted by Churchill's highest commanders to strike at the heart of Hitler's naval strength.

Three years into the war, the battle to keep Britain's vital sea lanes open was at its height. German battleships and U Boats were a lethal threat. In January 1942 the heaviest battleship ever built, *Tirpitz*, put to sea from Germany and steamed to the North Sea. She was a giant killing machine: 50,000 tons with eight 15-inch guns. Her speed of 30 knots and range of 8,000 miles made her a grave danger to Britain's Atlantic convoys. From her new base in a Norwegian fjord *Tirpitz* could break out into the Atlantic and cause havoc among cargo ships carrying precious munitions to the Soviet Union. This major ally of the West was struggling to survive after the German invasion of 1941.

There was only one dry dock on the Atlantic edge of Hitler's empire that could house *Tirpitz* or any other battleship in need of repair. This was the Normandie dry dock in St Nazaire at the mouth of the River Loire in German-occupied France. It was named after the great French liner *Normandie* and had been built large enough to accommodate her. *Tirpitz*'s sister ship *Bismarck* had been heading there for urgent repairs in 1941 when she was finally sunk by Swordfish aircraft from the British carrier *Ark Royal*.

By 1942 the destruction of the Normandie dry dock had risen

to the top of Britain's strategic imperatives. The RAF had made several attempts to obliterate it but without the precision weapons we have today was unable to bomb the gates successfully. It was decided that only a large force of Commandos on the ground could do the job. No one underestimated the scale of the risk involved in attacking one of the most heavily defended ports in German-occupied Europe. The best hope was, as the Director of Combined Operations Lord Mountbatten said, 'The fact that it is impossible will make it possible; the Germans will never believe we will do it.'

Intelligence about the layout of the docks at St Nazaire was partly provided by a very helpful shipyard worker who turned up at a French resistance base. He stripped off his clothes to reveal a plan of the docks wrapped round his torso. It was passed on to London, where Operation Chariot was being conceived at the highest level – an operation of breathtaking audacity. A force of 622 men would be carried in a flotilla of nineteen vessels led by a former American destroyer renamed *Campbeltown*. Her bow would be packed with a colossal charge of high explosive, and she would be rammed against the huge gate of the dry dock. The four-and-a-half-ton charge would be timed to explode once the crew were off and the operation was complete. The flotilla's Commando units would first attack the dry dock's gate-opening machinery and other targets in the dockyard. Each demolition squad would be accompanied by a 'protection team' armed with machine guns. Once the job was done, the men would assemble at an old pier in the harbour and be carried home in a fleet of waiting motor launches.

Lieutenant Purdon knew none of this as he reported for duty at Falmouth in south-west England in March 1942. He and the rest of the team just hoped they'd be going on what they called 'a show'. They were told always to wear duffel coats over their uniforms to avoid being spotted by German spies. Purdon and his four-man demolition team were housed in the wardroom of *Campbeltown* for the voyage to St Nazaire. They were told to sit on the floor because, as he later realised, 'they thought when *Campbeltown* hit the dry dock that we'd be knocked spinning.' Finally, as they neared the French coast they were given a detailed briefing on exactly what they were expected to do. One of Purdon's most experienced senior officers, Major Bill Copland, who knew more than most about the danger ahead, wrote a letter that he asked to be read to his wife if he didn't return: 'We sail in a day or two on a somewhat desperate venture, but one full of high purpose. If we succeed, and only the worst of ill luck will stop us, then we shall have struck a great blow for the cause of freedom. Remember too that if I do get blotted out I shall probably die in good company – for never did a finer crowd set out on a doughtier task.'

Campbeltown steamed off with the patrol boats and other escorts at 2 p.m. on 26 March. A day and a half later in the small hours of 28 March they were approaching the mouth of the Loire and St Nazaire from the south-west, Purdon and his men crouching in the wardroom. The destroyer had been reconfigured to look like a German warship and was flying the swastika flag in an effort to fool the multitude of German shore batteries on either side of the Loire estuary. Miraculously the *Campbeltown* managed to get within two miles of her target before the Germans suddenly realised they'd been duped. A blaze of searchlights swept to and fro and a devastating barrage of fire opened up on all the ships. Down came the German flags and up went British ensigns. 'All hell was let loose . . . they were coming at us at point blank range,' wrote one of the destroyer's crew. Down in the wardroom Purdon and his team had a close shave. They watched a huge red shell burst in through one wall and fly out the other – without exploding. 'The Germans were hitting the ship a hell of a lot,' said Purdon. 'On one occasion we hit the bottom. I remember thinking, I hope we bloody well don't get stuck here. We didn't, thank God.' The next thing he knew they'd crashed into the great dry dock gate at full speed. 'I remember poor old Cab Calloway who was one of my corporals. He was sitting under a small bookcase. I remember the look of surprise on his face when the bookcase landed on his head.'

Moments after *Campbeltown* hit the gate, Major Copland calmly walked into the wardroom and announced, 'Righto, boys, time we went now.' Purdon clambered off the deck and down a wooden ladder that two sailors were holding on the quayside. He looked around for his promised protection party. There was no sign of it. They'd got involved in a firefight and most had been killed. 'I decided what's the point of hanging around,' Purdon told us. 'No one's going to bloody well come, we'll do it on our own.' German bullets were flying everywhere and British casualties were mounting fast.

Purdon and his men knew exactly where to go. The RAF model of the harbour had been entirely accurate. They ran more than 300 metres along the west side of the giant dry dock and paused outside the door of the building that housed the winding machinery for the north dock gate. It was locked. Purdon fired his pistol at the lock. It had no effect except that the bullet ricocheted and nearly hit one of his team, Corporal Ron Chung, who remarked politely to Purdon that he'd come to France to be shot at by Hitler's men not by his own commander. Chung then used hammers to break the lock and they were in. It took them ten minutes to lay the charges. One corporal, Johnny Johnson, had been hit by a bullet and was in great pain but it didn't stop him packing in the charges. 'We knew exactly where to put them,' Purdon said to us. 'We'd already practised on a model of Southampton docks. It was plastic explosive – like plasticine in a way.'

The squad then had to wait while another team who'd been placing charges on the actual caisson gate of the dock completed their work under heavy gunfire. One was wounded but once they had got to safety Purdon could blow his charges. He pulled the pins of his igniters and 'the entire building seemed to rise several feet in the air . . . and then collapse like a house of cards.' Purdon and his men raced along the dockside to the point where they were supposed to meet the rest of the force. The plan was to fight their way there, then get on the motor launches that would take them home. But it soon became obvious that there was no way back to England. One glance at the chaotic scene on the river was enough. A huge concentration of German firepower had set alight nearly every one of the flimsy motor launches. These vessels had proved to be the Achilles heel of the operation. 'They were made of wood,' Purdon told us, 'and they had these bloody great gasoline tanks on board and they just went up.' Nearly all the launches were in flames or had disappeared. Purdon's commanding officer, Colonel Charles Newman, told him there was only one way out – to fight their way through St Nazaire and then walk 400 miles to Spain. The only good news he could offer was that the raid had been a success: the attacks on the winding houses and the caissons had put the great dock out of action. As long as the massive charge in *Campbeltown* exploded some time soon, the job would be complete.

What followed was an attempted breakout by a dwindling force of Commandos – all of them facing what seemed to Purdon to be 'the whole German army' determined to eliminate them. They had to charge across one bridge 'like a pack of rugger forwards'. Purdon had only a .45 pistol, which he called a popgun, but he was able to pump it into the slit of a pill-box on the far side of the bridge hoping to kill someone before he was hit himself. A grenade landed and burst near his feet. He was hit in the leg and shoulder and lifted clear off the ground. His battledress trouser was soon wet with blood but he found his leg still worked and 'I quickly forgot about it.' The British Commandos saw a German armoured car swing into a firing position 150 yards ahead, so they dodged left and there followed a frantic clamber across roofs and through gardens. One managed to grab a mouthful of breakfast as he raced through a house.

Dawn was breaking and Colonel Newman, who was still with Purdon and his group, suggested they find a cellar and attempt to hide out until nightfall. They found one that had a number of palliasses on which the wounded could get some rest. The only problem was that the house was directly opposite Gestapo headquarters, and minutes later German soldiers were shouting and stamping about on the floor above. Purdon was astonished that the Germans didn't simply lob a couple of grenades down the basement stairs. His colonel wisely chose to climb the stairs and tell the Germans he and his men would not fight a hopeless final battle. He was, observed Purdon, 'a kind-hearted man who decided "I am not going to have any more nonsense and bloodshed."'

The British captives were assembled in a café and one of the Commandos remembered hearing some Germans express admiration for what the British raiders had done. By now it was near noon and other prisoners were being brought in. One of them, Mickey Burn, was led by his captors along the quay past *Campbeltown*, which was still perched with her bow against the lock gates. He saw groups of Germans, soldiers and friends, gathered on the destroyer's deck apparently souvenir-hunting. The planned explosion had yet to take place. Burn prayed that the charge would be detonated – but not until he'd gone past.

It had been a long wait, but the *coup de grâce* finally came just as Purdon and others were beginning to believe the fuses had failed. By a satisfying coincidence a German was just saying with a sneer,

'If this is all you could do, was your raid really worth it?' when there was an almighty explosion. *Campbeltown* was blown open, taking the huge dock gate with her and leaving the dockside littered with the shattered remains of dozens of Germans. Purdon told us, 'I remember there was just a terrific cheer from all of us and the Germans got quite hysterical. One of them shouted, "We will shoot you." One of us shouted back, "Listen, you silly little man, either shoot us all or shut up."' No shots were fired but the prisoners were transported off. Purdon's wound was treated and 75 years later he could not forget the 'moronic naval youth who used my backside as a dartboard. When I was injected presumably with anti-tetanus, he threw the hypodermic syringe at my buttock, and I still recall the point of its needle hitting the bottom of my spine.'

Before they were moved to prison camp in Germany, Purdon and a group of British officers and men were escorted to the funeral and burial of their fallen comrades. 'It was', he said, 'a dreadfully depressing occasion but the Germans were correct and punctilious and obviously wanted to honour our brave men.'

Of the 622 who embarked on the raid, 168 were killed and 214 taken prisoner. Only just over a third made it home – five managed to escape to Spain, while the rest got away in the few motor launches that survived the heavy German gunfire. The human cost of the raid was high but it had achieved its purpose. The dry dock was not repaired until 1947, and *Tirpitz*, which might have done untold damage in the Atlantic, spent the rest of her life holed up in the Norwegian fjord unable to rely on repair in St Nazaire. The RAF finally bombed and destroyed *Tirpitz* in a massive raid on 12 November 1944.

Corran Purdon, determined not to remain a prisoner, managed to escape from the Nazi jail at Spangenberg Castle. But he was recaptured and, though he tried again, his new prison, Colditz Castle, proved too much for him. Since his release in 1945 he has revisited St Nazaire and the war cemetery several times. On one occasion the French Prime Minister told the veterans: 'You were the ones who first gave us hope.'

In spite of the high death toll, Purdon told us he felt honoured to have been on the St Nazaire raid. 'I'm just so bloody glad I went on it. That's why we joined the Commandos.'

SPIES AND INTELLIGENCE

SPYING IS AS old as warfare itself. The best military commanders from the dawn of time have sought intelligence about the land and the enemy ahead. Moses sent twelve men to 'spy out the land of Canaan'. They returned to say it flowed 'with milk and honey' but they also reported the Israelites faced formidable opposition if they tried to invade. The people were so scared that Moses kept them waiting another forty years to allow a new generation to lead them triumphantly into the Promised Land. Espionage can have a decisive effect.

Over the centuries the value of spies has been questioned but they proliferated nonetheless. Queen Elizabeth put considerable resources into the hands of her spymaster Sir Thomas Walsingham so that his agents could assure her security: they found out enough to seal the fate of Mary, Queen of Scots. Intelligence can win battles too. The Duke of Wellington believed that the information he gleaned from numerous British and Spanish spies helped him win the Peninsular War against Napoleon. When a Union soldier in the American Civil War found a copy of Confederate General Robert E. Lee's Special Order 191 detailing his campaign plan, General George McClellan was able to take action that fought Lee to a stalemate at the Battle of Antietam. The ingenuity of Alan Turing and others at Bletchley Park during the Second World War smashed the Enigma code, revealing German plans and helping pave the way to final victory.

By the time of the two World Wars and the revolution in modern technology, spying was an essential weapon of war – and peace. Spies themselves were skilled, courageous and highly resourceful operators, equipped with sophisticated means of communication and encryption of their messages. They were frequently exposed to great danger and risked execution if captured, even in the peaceful environment of the Cold War. Julius and Ethel Rosenberg were sent to the electric chair in 1953 for passing on American nuclear secrets, and Oleg Penkovsky, a Russian who gave Soviet missile secrets to the West, was executed by the Kremlin in 1963. And one of the ironies is that for the most part spies don't win acclaim for their successes: in the lonely world of espionage you're only famous if you're caught. Most of the world's greatest spies remain unrecognised.

We've chosen to tell the stories of three who have been recognised, each for very different reasons. Ursula Graham Bower was the eyes and ears of British resistance to the Japanese advance on India in the 1940s. One of the reasons her story appeals to us is the arresting quality of her own account of her adventures. Our other Ursula is one of the most skilful Soviet spies of the Second World War. Born Ursula Kuczynski, she had many aliases but the main one was Sonya. She groomed one of the most notorious British spies to hand over nuclear secrets to the Soviet Union. But our choice of spies begins with the man still regarded as one of the most infamous American traitors. He is certainly one of the world's greatest turncoats, General Benedict Arnold. His conspiracy was as ambitious as any in military history.

10

Benedict Arnold

America's Greatest Turncoat, US War of Independence, 1775–81

FEW PEOPLE BELIEVED it could be done. In the late autumn of 1775, half a year into their struggle for independence from Britain, 600 hardy American troops stood on the banks of the St Lawrence River. Inspired by their resourceful and intrepid leader, they had struggled through the wilderness of Maine. Their march through some of the world's most inhospitable territory in order to surprise the British in their colony of Quebec caused astonishment on both sides in the American War of Independence. Their commander was to become America's most single-minded and aggressive soldier, Colonel Benedict Arnold. In his determination to achieve his goal he had led his men across raging torrents and through mountain passes in appalling conditions. 'The march of that little army,' wrote one delegate to breakaway America's Continental Congress, 'is thought to be equal to Hannibal's over the Alps.' George Washington, the commander-in-chief, commended Arnold on his 'enterprising and persevering spirit – it is not in the power of any man to command success, but you have done more, you have earned it.'

Benedict Arnold was in almost everyone's judgement the most dynamic officer in the army and robustly devoted to the cause of the emerging United States. And yet this was the man who five years into the war would betray his young country in an astonishing act of treachery. He changed sides, and offered to hand over one of America's key garrisons and even General Washington himself to the British army. By fighting for both sides in the War of Independence Benedict Arnold earned himself the title of America's greatest traitor. And the story of how he and his enchanting wife, Peggy, schemed with the British through the agency of the debonair British spy John André is as full of murky skulduggery as any in the history of warfare.

Little in his background suggested he would become infamous. Born in 1741 to a well-to-do family in Connecticut, young Benedict was worried he might grow up to be a coward. He decided to put himself to the test. While his playmates watched aghast, he threw himself on to a mill wheel and clung to it for dear life as it revolved in the turbulent millstream. After a few turns the boy swam off triumphant. He was to go on to display outstanding bravery and to expect it in others. A bit of a ruffian, he enjoyed a good scrap.

Arnold was short, muscular, with a prominent pointed nose and blue eyes. He had unbounded energy and it was not politics but fighting that he enjoyed. As an officer in the Connecticut and then Massachusetts militia he was a formidable commander with a preference for attack rather than defence. At the beginning of the war with Britain his aggressive leadership made him a natural choice as leader of the force dispatched to capture Fort Ticonderoga, a key British base at the foot of Lake Champlain. He did just that, and the fort's stock of arms and ammunition gave a useful boost to the growing American rebellion in May 1775.

Arnold's pugnacity, so valuable in battle, was matched by his determination to be in charge. It was to earn him many enemies in the army and in public life, especially among those jealous of his success. When, after his triumphant raid on Fort Ticonderoga, he learned that another colonel had been appointed commander of his force as it moved on northwards towards Quebec, he exclaimed, 'I will be second in command to no person whatsoever,' and resigned his commission. He also brusquely demanded that the state refund him all the expenses he'd incurred on supplies for his attack on the fort. Out of a job and at odds with his superiors, he suffered a further shock when he reached home and found his wife had died. He and Margaret Mansfield had been married only eight years. He now had three young children to look after.

But his career took a new turn when he met George Washington, the newly appointed American army commander, in the late summer of 1775. They immediately hit it off. Washington recognised Arnold's leadership qualities, re-engaged him and sent him off on his famous march to Quebec. Although he conducted a blistering assault at the head of his men through the lower town, he was seriously wounded in the thigh and failed to capture the city. In spite of this reverse, Washington admired Arnold's inspired leadership and his tactical skills, and when the Americans pulled out of Canada in 1776, he made him commander of the Lake Champlain area. As the British moved in, Arnold conducted an impressive fighting withdrawal against heavy odds.

He again lost his temper when Congress drew up a list of charges against him: he'd allowed smallpox to rage among his soldiers, it was alleged; he had not provided them with enough supplies; and, worst of all, he had been responsible for the failure in Canada. He was also shocked not to be promoted to major-general. He wrote a letter of complaint to Washington, demanding a court of enquiry into his conduct and talking of the 'ingratitude' of his countrymen. Washington was in two minds about Arnold. He was alarmed by his fits of paranoia but anxious to retain the services of an officer he greatly admired. He wrote back in warm terms but added: 'I cannot see upon what ground you can demand a court of enquiry.'

On 25 April 1777 Arnold was at home about to write another letter of resignation when there was a loud knocking on his door. An army messenger told him he was badly needed back in the front line. The British had invaded Connecticut. He was soon reunited with the troops he loved commanding and successfully drove off the British in a series of engagements. On one occasion, surrounded by British troops, his horse was shot under him and he fell face down in the mud.

'You are my prisoner,' cried an enemy soldier standing over him with a bayonet.

'Not yet!' retorted Arnold, his pistol still in his hand. He pulled the trigger, mortally wounding his assailant. Dodging the falling body with its flailing bayonet, he raced through a volley of musket fire into the cover of a nearby wood.

Washington was delighted at Arnold's dramatic escape and continuing success and Congress awarded him the promotion to major-general he'd been seeking. But Arnold was angry again that others had been appointed major-generals when they were actually junior to him. He demanded that his seniority be recognised. He was also beginning to sense disagreement among his political bosses – the leaders of the American Revolution – on a point of principle. There were those who believed that America's finest soldiers like Arnold should be celebrated as war heroes, but others who feared such popular warriors could be a threat to the new republic's immature democracy. Arnold faced the unpalatable possibility that his heroism was making him unpopular.

By July Arnold was once again on the point of writing another letter of resignation when a major British invasion led by General John Burgoyne advanced south from Canada in order to join up with the important British stronghold in New York City. Once more Congress asked Arnold to come to the rescue. Unable to resist a military challenge, he swallowed his resentment and agreed to serve under General Gates who'd been a friend earlier in the war. But it wasn't long before Arnold's impatience at the pace of this new campaign got the better of him. During the fighting around Saratoga – where Burgoyne was held up halfway down the road to New York – Arnold was told by Gates, probably out of jealousy, to stay away from the front as a new battle began on 7 October 1777. Arnold was then appalled to see that far from rushing to the front himself, Gates remained in his tent. Unable to contain himself, Arnold leapt on his horse and charged off to the sound of the guns. Gates immediately dispatched an aide to recall his disobedient subordinate but Arnold was too quick and was soon in the midst of the fray, waving his sword above his head, shouting orders and encouragement to the men he knew so well. Time and again he rallied the troops and led charges against the British. The battle was won and ten days later Burgoyne surrendered his army at Saratoga. It was one of the most crushing British defeats in the war. But it was the last time Arnold fought for the Americans. He had been wounded in the thigh, even more severely than he had been at Quebec. It would take him months to recover.

Our story now moves to the city of Philadelphia where the Declaration of Independence had been signed in 1776. It was here that fate was preparing a surprise for Arnold. The British had captured the city in the autumn of 1777. For some Philadelphian residents, like the well-connected Shippen family, the return of the British was no great disappointment. The Shippens had always been careful to declare themselves neutral in the war, but they privately believed their interests were best served by staying loyal to the king in London. Edward Shippen, a lawyer, had never recognised the revolution as legal. For his ravishing daughter, Peggy, it wasn't the law that mattered as much as the welcome jollity, elegance and flattery that British officers brought with them. Peggy, a vibrant partygoer, was swept off her feet by the heady flavour of George III's London that the smart young soldiers and sailors brought to Philadelphia's society.

One of those soldiers was the handsome John André, a twenty-seven-year-old charmer, whom Peggy had already met when he'd been a prisoner of the Americans three years earlier. André was poorly treated during his captivity and his ordeal made him more determined than ever to crush the American rebellion. He was an aide to the new British commander in New York, General Sir Henry Clinton. But for the moment here was a chance to renew his dalliance with this delightful young beauty: just how far their relationship went we can only judge from the fact that Peggy kept a lock of André's hair for the rest of her life. What neither of them knew yet was that they'd both be partners in one of the greatest conspiracies in US history.

In June 1778, Arnold, still hobbling around on a stick after his injury at Saratoga, had a lucky break. Washington's army reoccupied Philadelphia and the commander-in-chief honoured his old comrade by awarding him the military command of the city. Arnold took up residence in the magnificent mansion where the British governor had once hosted lavish parties. Since he was not well enough to fight, he decided to make the best of life in his new domain. In the hectic new social whirl that soon enveloped upper-class Philadelphians, it wasn't long before Arnold met the enchanting Peggy. She was as full of passion and mischief as he was. They were at once enthralled by each other and married in the spring of 1779.

Arnold soon started abusing his authority by engaging in a number of dubious activities. He used his position and his contacts on both sides of the conflict to make money. For example, he promised New York merchants under British rule that he would protect their property once the Americans took over as long as they paid him a fee in advance. His illicit activities were seized upon by those who disliked him and they brought charges against him. He was finally convicted but only of two minor military charges and sentenced to receive a 'reprimand' from George Washington. But the pressure of first being under suspicion and then not being fully acquitted hung heavy on Arnold and left him again complaining of the 'ingratitude' of his countrymen.

Scholars debate to this day what it was that turned Benedict Arnold against his country and led to the sensational conspiracy of 1780. There is little doubt that Peggy did her bit to urge him to change sides, using her association with John André to negotiate an extraordinary deal between Arnold and the British commander-in-chief, General Sir Henry Clinton. Peggy worked on her husband's resentment and also on his quiet conviction that Britain might well win the war and that with his help, Britain and the loyalists could return America to a state of happiness, peace and prosperity for families like the Arnolds and the Shippens. Whatever his reasons, Arnold was about to commit an act of treachery against his country and betray his commander-in-chief. It was an unforgivable action by someone who had received nothing but friendship from George Washington.

The British spy John André was delighted to play a pivotal role in this plot. He was now Clinton's adjutant general, a senior staffer and in charge of British intelligence with the particular job of encouraging disaffection among the American rebels. Besides, he badly wanted his temporary rank of major to be confirmed – what better way than to persuade Arnold, one of the top enemy generals, to change sides? Through a trusted intermediary he and Arnold communicated with each other in coded letters. Each side had access to the same dictionary and could decode messages by referring the numbers in the letters to pages and columns in the book. The deal with the British was that Arnold would persuade Washington to appoint him commandant of West Point, a vital strongpoint on the Hudson River, later the site of the US military academy. Arnold would then ensure that West Point was so poorly defended that Clinton's British forces could easily capture it.

That alone might have been enough to allow Britain to win a decisive advantage in the war. But there was more. Arnold knew enough detail about George Washington's movements to be able to promise Clinton that if the British struck at the right moment they would be able to seize the US commander-in-chief as well. Clinton knew that would bring the United States to its knees. As the contacts with Arnold grew in intensity through the summer of 1780, Clinton prepared his forces for the *coup de grâce* and warned his colleague Admiral Rodney to prepare his Royal Navy fleet to sail up the Hudson.

Arnold constantly badgered George Washington for the West Point job and was under the impression that he had got it. Anxious to confirm this, he rode up to his commander-in-chief near West Point one hot midsummer's day. Wreathed in smiles, he asked whether Washington had 'thought of anything for him'. Washington later remembered the exchange in detail. He told Arnold he had indeed a 'post of honour' for him. But when he told him what that post was, Arnold's face dropped dramatically. Washington offered him not the job at West Point but the command of the entire left wing of his army. And, recalled Washington, 'instead of thanking me or expressing any pleasure . . . Arnold never opened his mouth.' Washington could hardly believe it when, as they rode off together, Arnold stressed how lame his leg was and begged Washington to let him have West Point as a 'granny's post'. Washington suspected nothing of the man he still treated as a friend. And three days later he granted Arnold the job for which he was massively overqualified but which precisely suited his purpose.

The new commandant of West Point made his residence in a fine mansion called Beverley just across the river from the fortress. He also enlisted a man whom few people trusted as his henchman – a shady lawyer called Joshua Smith, who was naturally keen to be on the best of terms with the new grandee at West Point. When Arnold and André concluded that the only way to seal their deal was to meet secretly, Smith seemed to Arnold the ideal man to arrange the meeting. The first attempt to meet on the riverbank on 11 September had to be aborted when Arnold's small boat narrowly escaped being sunk by shellfire from British gunboats. André had forgotten to tell them to hold off.

Another date was set, 20 September. In the meantime, to Arnold's delight, Peggy and the Arnolds' first baby, Neddy, had moved into the new house. So Peggy was there to witness the drama of the meeting between her husband and André and the debacle that followed. It was arranged that a British ship, HMS *Vulture*, would bring André up the Hudson River. Joshua Smith would then row to the warship at night and ferry the waiting André to the river bank for a meeting with Arnold. That meeting would finalise the storming of West Point and the capture of George Washington. The plan was to seize Washington when he came to the Arnolds' house for a meal in four days' time. Peggy would prepare a major feast which she would make last as long as possible in order to give the British plenty of opportunity to capture the American leader.

Clinton, the British commander, approved the plan but gave André three warnings. He was told not to wear a disguise, to meet Arnold only in no-man's-land between the warring armies' front lines, and not to carry any incriminating papers that would reveal him as a spy. André would disregard all three of Clinton's warnings in the murky events that followed.

Arnold gave his sidekick Smith a pass for both him and André to move safely through American lines. As the sun went down on the appointed day Arnold rode to Smith's house and found him arguing with his boatman, who didn't want to risk rowing to the British ship. It took the two men much of the night and several inducements including a stiff whisky to persuade the boatman to take Smith across the water. Once alongside *Vulture*, Smith met the warship's captain and was introduced to André who was operating under the pseudonym of 'John Anderson'. Ignoring Clinton's warning that disguise would put him in danger of being shot if he was caught, 'Anderson' donned a plain overcoat with high collar over his red uniform and stepped into the boat. When they reached the bank Arnold appeared in the pitch darkness and disappeared into a grove of trees along with André. It was the first and only time the men were to meet.

There's no full record of what they discussed but Arnold later said he was promised £6,000 of the £10,000 he asked for in return for his information and for the risk he was taking. He no doubt told André how West Point was defended and what would be needed to capture it as well as the timing of his and Peggy's plans to host General Washington on the coming weekend. As dawn approached Smith interrupted them and said it was time for 'Anderson' to return to *Vulture*. But the boatman now flatly refused to row back, and Arnold suggested to André that they retire to Smith's nearby house. It was there that they heard the roar of artillery and watched American guns fire at *Vulture* even though Arnold had instructed his men not to target the British warship. *Vulture* disappeared downstream. Smith recalled seeing André's face drop and 'the energy with which he expressed his wish to be on board'. Desperate to get back to British lines André took Arnold's advice to ride off with Smith. He now ignored the most important of Clinton's warnings. He agreed to carry a whole sheaf of papers, which Arnold made him conceal in his stocking – papers that contained the plans to betray West Point. If he were caught the documents would be blatantly incriminating: many were written in Arnold's own hand. Confident that André would make it back to the British side and that the raid on West Point would soon be under way, Arnold headed home to Peggy to prepare for Washington's visit.

But Smith left André well short of the British lines, and when Clinton's spymaster ran into three armed men he mistook them for British skirmishers. 'Thank God I am once again among friends,' André said with a confident smile. 'I am glad to see you. I am an officer in the British service.'

'We are Americans,' growled one of the three to André's horror, and even though he produced the pass Arnold had given him, which should have allowed him through American lines, they insisted on searching him.

He was soon stark naked with just his stockings and boots on, and when his captors insisted he remove those too, the telltale documents fell out. As André was led away on horseback with his hands tied behind his back, he told the men, 'I would to God you had blown my brains out when you stopped me!'

When they reached the American lines, the men handed over André and the papers to a senior officer who noticed that the documents with their information about the defences of West Point were written in the same handwriting as the pass Arnold had given André. The officer was puzzled. It looked as if Benedict Arnold had written the incriminating documents but there was no firm proof of his guilt. He ordered the papers to be taken to George Washington along with news of André's arrest. But his sense of duty now led him to make a terrible mistake. He decided it was only right also to inform Arnold, who was after all his commanding officer. It was a decision that was to save Arnold's life because Arnold received the message minutes before Washington received his.

Because of several delays, neither message had been delivered by the 24th when the Arnolds were expecting Washington to the fateful meal. Peggy was upstairs with baby Neddy, Arnold waiting to greet his c-in-c. Suddenly in burst a messenger bringing Arnold the spine-chilling news that a 'John Anderson' had been arrested carrying what looked like incriminating papers, which had now been sent to General Washington. Arnold, realising that the game was up, shot out of the room in a flash and rushed upstairs to whisper frantically to Peggy that all was lost. With a last goodbye to her and a hug for Neddy, he raced down, leapt on his horse, cantered to the river and ordered his boatmen to row down the river for all they were worth.

Minutes after Arnold made his escape, Washington arrived and was shocked to find Peggy in floods of tears. Whether she was uncontrollably distraught at Arnold's flight or feigning a fit of hysterics to deceive Washington, her behaviour bought her fugitive husband time to make good his escape. It also later persuaded Washington that she was innocent. He soon received the message containing news of André's arrest and the discovery of the documents revealing the enormity of Arnold's treachery. 'Arnold has betrayed us,' he exclaimed. 'Whom can we trust now?'

André, the British spy, was tried and briskly convicted. Little more than a week after he was captured his request to be shot by a firing squad was refused, and he was made to stand on a wagon with a noose around his neck. Allowed to speak before his hanging he said, 'I pray you to bear witness that I meet my fate like a brave man.' The wagon was dragged away and André was left hanging lifeless.

Arnold and Clinton, who were now together in British headquarters in New York, expressed their dismay and outrage at André's execution. 'Washington,' said Clinton, 'must answer for the dreadful consequences.' But they were unable to avenge André's death. Arnold served as a British general for the rest of the war; he had little time to have much impact on the fighting. A year after the failed plot, the British surrendered at Yorktown and the war was effectively over. America had won its independence, and Benedict Arnold, who might have enjoyed a place of some honour in the new republic if he had stayed true to it, spent the rest of his life – with Peggy and two more children – restlessly seeking a home in Britain and Canada. He died in 1801 reviled by Americans and with few admirers in the rest of the world.

The body of the British spy John André was handed over by the USA forty years later. It lies in Westminster Abbey near a plaque that reads: 'he fell a sacrifice to his zeal for his King and country . . . universally beloved and esteemed by the army in which he served.'

11

Sonya

Spying for China and Russia, 1930–50

To all appearances, the couple walking arm-in-arm through the Oxford countryside looked perfectly normal. As they strolled they talked of movies, books, plays and concerts. He wore glasses and had a modest air. She had dark hair, knowing eyes, a wide smile and sometimes brought a bicycle with her young daughter's wicker seat on the back. No one would have guessed the man was a nuclear scientist who was passing top-secret information about the atomic bomb to a woman described as 'one of the top spies ever produced by the Soviet Union'.

The year was 1942 and Sonya, originally from Berlin, was living in Oxford with her British husband and three children. She told people she was 'just a housewife' but she had been spying for the Soviets for twelve years. The man she met on her country walks was Klaus Fuchs. He too was a German immigrant who'd fled his homeland after Hitler came to power in 1933. A talented physicist, he was working with a team of scientists in Birmingham on the British atomic bomb research project. Both Sonya and Fuchs were committed Communists who believed that helping Russia would lead to a better world. The information he passed on through Sonya is thought to have accelerated Russia's atomic weapons programme by many years.

The name Sonya was a cover, just one of many aliases she adopted in a life where she was constantly exposed to danger. Her real name was Ursula Ruth Kuczynski and she was born in Berlin in 1907. One of six children, she had a privileged upbringing but, from an early age, was shocked at the gulf between rich and poor. Things in Germany were to get dramatically worse after the First World War, both economically and politically.

The country of Sonya's youth was riven by economic and political turmoil. After losing the First World War, Germany had been forced to accept a huge reparations bill which helped to contribute to the disastrous inflation of 1923. At the start of the war, in 1914, there were four German marks to the dollar. By 1921 it took seventy-five marks to buy one dollar. At the end of 1923, a dollar was worth 4.2 trillion marks. Political tensions also ran high with left-wing parties pitted against right. The Nazi Party of Adolf Hitler, which blamed the ruling class, Communists, trade unions, liberals and Jews, among others, for Germany's woes, grew in popularity during the 1920s. As did the Communists who promised a better society.

Appalled at the rise of Fascism and concerned about the future, Sonya joined the Communist Party of Germany. She was just nineteen years old. 'I observed the wealth of small, privileged circles and the poverty in which so many people lived; I saw the unemployed begging on street corners. I pondered over the injustices of this world and how they might be eliminated,' she wrote in her autobiography.

Sonya entered the murky world of espionage in the early 1930s. By now she was married to architect Rolf Hamburger, not himself a Communist. When he was offered a job in China, he leapt at the opportunity. This was Nationalist China well before the Communists took over with a glaring gap between rich and poor. From the moment Sonya's ship arrived in booming Shanghai, she was appalled by the poverty and inhumane treatment she witnessed. It was, she wrote, the worst she had ever seen.

> Porters, emerging from the bowels of the ship and padding the steep planks to the quay, followed each other so closely that their heavily laden bamboo poles almost touched. Sweat streamed down their bodies; thick veins protruded from necks, temples and legs . . . Encircling the ship in floating tubs were beggars, moaning cripples with stumps for arms and legs, children with festering wounds, some blind, some with hairless scab-encrusted heads.

Sonya decided to take action. The Communists in China, soon to be led by Mao Zedong, were under attack from the Nationalist government. This struggle gave Sonya the perfect opportunity to spy for a cause she believed in. She met the handsome, charismatic Richard Sorge, an intelligence officer employed by the Soviet Union to help their Communist Chinese comrades. He later became one of the most famous spies of the Second World War. His cover in Shanghai was as a journalist working for a German newspaper. When Sorge asked to use the Hamburgers' house to hold meetings with Chinese Communists, Sonya agreed. She also allowed him to store weapons and suitcases full of secret documents in her cupboards – all without telling her husband. For three years weekly meetings between Sorge and Chinese Communists took place under her roof, a crime punishable by death.

Sorge asked Sonya to go to Communist demonstrations and report back on what she saw. On one occasion she stood outside a Shanghai department store, her arms full of purchases to look inconspicuous. She watched as Communists were beaten and arrested and became more committed than ever. 'In many cases arrest meant the end. I looked into the faces of young revolutionaries whose death sentence had been pronounced at that moment, and I knew – if only for their sake – that I would carry out any task asked of me.'

Astonishingly, Sonya's husband remained unaware of what she was doing. It was only when she was asked to hide a Chinese Communist whose life was in danger that she finally told Rolf what she was up to. Surprised and upset, he begged her not to take such risks but in the end he 'endured the difficult circumstances created by my work in the hope of saving our marriage'. Sonya's work was noted by Sovet intelligence agents. They were impressed, although one Russian agent complained that she should look 'more ladylike' and 'wear a hat'. In 1933 she was invited to attend a six-month training course in Moscow. That would mean leaving Rolf and their two-year-old son, Michael, but she accepted with Rolf's support.

'A course of instruction,' she told him, 'would help me to do my work better.' Her mission was to fight the enemies of Communism wherever they were. At the headquarters of the Soviet Intelligence Agency, known as Moscow Centre, she was given the cover name Sonya. Sonya means 'dormouse' in Russian and it was also the title of a popular song at the time. Fittingly, the lyrics were 'Sonya dances to Russian tunes'. She trained as a radio operator. She built transmitters and receivers, took Russian language lessons and attended political lectures. Her first posting was in 1934 to Mukden in Manchuria which the Japanese had invaded three years earlier. She insisted on taking her young son with her. Her job was to establish contact with Chinese groups who attacked Japanese commando posts and bombed railway lines. She used a transmitter to pass on reports and questions from the partisans, feeding back to them advice from the Soviet Union.

Her transmitter was smuggled in from Shanghai in the bottom of a large armchair. Hiding it proved challenging. It ended up in a large chest under Sonya's desk covered with moth-balled winter clothes in the summer and lightweight outfits in the winter. When new valves were needed for the transmitter, she travelled to China with her son and sewed them into Michael's teddy bear to avoid detection at the Manchurian border checkpoint.

One of her most important tasks during this time was to provide resistance fighters with explosive materials. With little Michael in tow 'to make it all look as harmless as possible', Sonya would visit chemists and buy the necessary chemicals. But when one dissident was arrested by the Japanese, some of Sonya's chemicals were discovered at his house. Moscow Centre ordered her to leave Mukden immediately and move to Peking.

It's no surprise that Sonya's work took its toll on her marriage. Her husband continued working in Shanghai while she was in Mukden and Peking. And to complicate matters, she had an affair with a fellow spy and became pregnant. Far from worrying about having a second child by another man, Sonya boasted 'a baby would provide excellent legal cover'. The long-suffering Rolf, who was devoted to his wife, did everything possible to save the marriage. He offered to bring up the new baby as his own. He even joined the Communist Party so he too could work as a spy.

In 1936 Moscow Centre sent the couple to Warsaw. The Soviets were anxious to keep an eye on the anti-Communist Polish government which had signed a treaty with Hitler. Sonya made contact with clandestine Communists and transmitted their reports back to Moscow. Sonya's second child, Janina, arrived but that didn't stop her from moving with her children 400 kilometres to Danzig (present-day Gdansk). She learned Polish, gathered information about the German U-boats being built there and gave advice to a group preparing guerrilla attacks against the expected Nazi invasion.

Transmitting reports about Poland to Moscow proved dangerous and challenging. In Danzig a week after moving into a rented house, Sonya discovered that a nearby electric power station was causing a loud humming noise that caused problems with her signal. Even though she feared that leaving so quickly could look suspicious, she and her two children moved into an apartment. Her neighbour was a Nazi official whose wife loved to gossip. One day she asked Sonya whether she got interference on her radio at night. She said her husband suspected someone was sending secret transmissions and had arranged for the whole apartment building to be surrounded when he returned from a business trip. Sonya promptly packed up her transmitter and told Moscow she could no longer send messages from that location.

After finding a new place to send and receive messages she was surprised to get a message from Moscow announcing that she had been awarded the Order of the Red Banner, the top Russian military award for extraordinary heroism, courage and dedication. She modestly thought the message was meant for someone else.

Recalled for more training in Moscow in 1937 and 1938, Sonya was treated as a hero. She went to the Kremlin to receive her Order of the Red Banner medal from none other than Mikhail Kalinin who went on to become chairman of the Supreme Soviet. She was made a major and later a colonel in the Red Army.

During her Moscow stays, Sonya attended a school that trained intelligence officers and partisans. She learned to make a new, more complicated transmitter and was taught how to turn chemicals into explosives, how to cut detonating wire, set contacts, ignite fuses and build a clockwork mechanism that would explode when trains went past.

When, in 1938, the Russians offered her a choice between spying in Finland or Switzerland, she chose Switzerland because it would be easier for her to make contacts there. Even though it was clear to Sonya that their marriage was effectively over, her ever faithful husband Rolf loyally accompanied her and the children to the Swiss Alps. They found a small farmhouse in the mountains high above Lake Geneva and built a transmitter together. Even though Switzerland was neutral then and during the Second World War, amateur radio communications were forbidden. At first the transmitter was hidden under the floor of a linen cupboard, then moved to a hole in the ground. It ended up in a coal shed. A hayloft attached to the house proved an ideal place to hide what Sonya called 'illegal materials'.

Sonya's 'idyllic little house' in the mountains and her cover as a mother of two meant she could work very effectively in Switzerland. She made contact with people in the League of Nations, headquartered in Geneva, and sent their views about the German war effort back to Moscow. She also recruited agents to send on sabotage missions into Germany. One of these was Len Beurton, a Communist who'd fought with the anti-Fascist British Brigade in the Spanish Civil War. At Sonya's command, he travelled to Germany with an incendiary device that fitted into a cigarette case to blow up a Zeppelin airship. While in Munich in early 1939, Len ate in an inconspicuous restaurant. Sitting next to him was an attractive brunette waiting for someone. Her name was Eva Braun and her dinner date turned out to be her lover, Adolf Hitler. Len discovered that the restaurant owner had supported Hitler long before the Nazis came to power so the Führer was a loyal customer. Sonya reported this unexpected discovery to Moscow Centre and pointed out that the restaurant would be an ideal place to assassinate the German leader. She never received a response to her bold suggestion.

On 23 August 1939 the Soviet Union signed a non-aggression

pact with Germany. Sonya admits she was 'shocked and bewildered' that her Communist heroes could have anything to do with a Fascist like Hitler, but after long discussions with fellow Communists she came up with this rationalisation: 'We realised that the western powers wanted the communist Soviet Union and Nazi Germany to destroy each other, in which case they would emerge from the brawl laughing. A stop had to be put to that little plan.'

Len Beurton proved a useful and convenient way for the Russians to give Sonya her most important posting of all. He had a British passport. Her handlers in Moscow suggested she divorce Rolf and marry Beurton. This would allow them to shift one of their most valuable spies to England. Rolf, ever the gentleman, agreed to the plan. Sonya married Beurton in February 1940, five months after the start of the Second World War. It may have been an arranged marriage but Sonya and Len were to have a long and very happy life together.

Re-named Ruth Beurton, Sonya arrived in Oxford in 1941 with her two children, Michael and Janina. Len remained behind to continue spying in Switzerland. By this time, Hitler had invaded Russia and the Soviet Union had changed sides. It was the friend of the west, a vital ally against Hitler. But Sonya was highly critical of the British government which she believed was withholding important military information from the Soviets. Moscow Central ordered her to meet a contact in central London. She was given what she described as 'recognition signs and passwords'. Told to infiltrate political circles to discover whether Britain was planning to 'come to an understanding with Hitler', Sonya also cultivated sources in the British military. A Royal Air Force technician, eager to help the Soviet Union, provided blueprints of new British fighter planes. Another contact handed over information about tank-landing operations and supplied a piece of equipment used for submarine radar.

Sonya transmitted the information she gathered late at night while her children slept. Years later Janina claimed she had no idea that her mother was a spy although she 'often wondered why my mother slept during the day. Other mothers did not do that.' Michael too said he was unaware of Sonya's activities although he realised something unusual was going on. 'She was meeting with these Communists who came fairly often. Whenever they came, they shut the door and I had to go out or go to bed.'

Sonya did not find it difficult to find sympathetic moles. Many, like her, believed the British and Americans took too long to open up a second front against Hitler, leaving the Soviet Union to 'fight and bleed – practically on her own – for a long time'. The Soviet Union, she believed, was 'the allied country that was fighting the hardest and bearing the heaviest sacrifices in the war against fascism'.

Sonya's biggest coup was gaining the trust of atomic scientist Klaus Fuchs. She learned from her brother Jurgen, a Communist economist who taught at the London School of Economics, that 'a comrade with worthwhile military information had lost touch with the Soviet Union for quite some time' and wanted to re-establish relations. She transmitted this information to Moscow and was ordered to make immediate contact with Fuchs.

For two years, Sonya and Fuchs met in the countryside near Oxford, never for more than thirty minutes at a time. Ironically, her route to meet Fuchs took her by Blenheim Palace. She had no idea that it was the headquarters of MI5, Britain's security service. They left messages for each other in a 'letterbox', a special place Sonya dug out between the roots of a tree. Sonya described Fuchs as 'calm, thoughtful, tactful and cultured' and, as she gained his confidence, he started passing her information. One day Fuchs handed her a thick book of blueprints with complicated formulae and drawings. They were top-secret copies of the work being done on Britain's atomic bomb. He asked her to forward them to Moscow.

When, in 1943, Fuchs moved to America to work on the Manhattan Project (a joint American, British, Canadian operation that produced the first atomic bombs), Sonya set up a Soviet contact for him in New York. This allowed Fuchs to send the Soviets information from New York and from the Los Alamos laboratory in Nevada that manufactured the nuclear bombs dropped on the Japanese cities of Hiroshima and Nagasaki at the end of the Second World War. Incredibly, and to us unconvincingly, Sonya always claimed she was unaware that Fuchs was passing on secrets that would lead to the atomic bomb.

After Fuchs left for America, Sonya continued working her contacts and sending information to Moscow Centre. She was deeply flattered when her Soviet boss sent a message of congratulation saying, 'If we had five Sonyas in England, the war would be over sooner.' In her view, his words justified what she was working for: 'the destruction of fascism and an end to the war for the sake of all humanity'.

Sonya's 'war effort' came to a sudden end in the summer of 1946. On the eve of the Cold War between East and West, just when she might have expected to be useful, Moscow Centre unexpectedly broke off contact with her. Perhaps it was because the Soviets had learned that Sonya was being watched by British intelligence. She was indeed questioned by 'two gentlemen' who turned up at her house. They were from MI5. Their report described Sonya as 'unimpressive with frowsy, unkempt hair'. When they informed her that a Soviet defector admitted to working with Sonya and Len in Switzerland she refused to answer questions. They left and their report on the Beurtons states: 'There is no reason to suspect them of present or even recent espionage activities. It is not anticipated that any further steps will be taken.' It was a close shave.

Sonya's life was very different once she stopped spying. Without the financial security of regular payments from Moscow Centre, she was forced to give German lessons and to rent out rooms in the house she shared with Len, who'd moved back to England, and their three children (Peter, Len's son, was born in 1943). Their quiet life came to an abrupt end in 1950 when Sonya saw a newspaper headline saying Klaus Fuchs had been arrested in Britain. After being interrogated by MI5, he had confessed to being a spy. The newspaper report mentioned Fuchs had met with a 'foreign woman with black hair' in a town near Oxford. Sonya buried her transmitter and on 28 February, the day before Fuchs's trial started, she and her two younger children flew to West Berlin, crossed the border and found sanctuary in Communist East Germany. Her son Michael later recalled that his mother seemed absolutely normal as she left their Oxford home. She told him and their neighbours she was going on a short holiday. Len joined the family in East Germany after a few months. Michael stayed in Britain another year to finish his studies at Aberdeen University. Sonya hadn't left a moment too soon. Her accomplice,

Klaus Fuchs, was found guilty and sentenced to fourteen years for spying.

Sonya worked in various jobs for the East German government but her real passion was writing. Under the name Ruth Werner she produced children's books. Her own children remained unaware of her secret life until one day she invited them to tea and announced she'd been a Russian spy. She came clean to the wider world with the publication in 1977 of her autobiography, *Sonya's Report*, the story of her life as a spy. It became a bestseller in the Communist world. Sonya's children survived their unusual childhood surprisingly well. Michael became a well-regarded Shakespearean scholar in Germany. Janina was a teacher and Peter a biologist and philosopher at the Academy of Sciences in East Berlin. Sonya's husband Len died in 1990. He and Sonya had been married for fifty-seven years. Her first husband Rolf Hamburger spied for the Soviets until his arrest in Iran in 1943. He spent five years in an Iranian forced labour camp before being sent to Russia. He moved to East Germany in 1955 and practised architecture in Dresden where he died in 1980. Sonya herself lived on and died in reunited Berlin in 2000, at the age of ninety-three. She never doubted that her mission was to make the world a better place and never apologised for being a Soviet spy. But the experience clearly stayed with her for ever. 'If a nightmare haunts my sleep, the enemy is at my heels and I have no time to destroy the information. If I find myself in new surroundings, I am forever looking for hiding places for illegal material.'

12

Ursula Graham Bower

Guerrilla Queen, Nagaland, 1942–5

'I KNEW HOW TO live in the jungle,' wrote the beautiful English debutante who became the first female guerrilla leader in the history of the British army. One of our more unlikely heroes, she played a modest but vital role in the defeat of the Japanese advance on India.

Born in England in 1914, Ursula Graham Bower was educated at the posh private boarding school Roedean. She had hoped to study anthropology at Oxford University but when her parents divorced and money was tight, Bower was taken out of school in favour of sending her younger brother to university. 'I heard it said that girls always marry. It was no good spending money on me because I wouldn't want a career – while my brother would,' she said later in an interview.

Typical of young ladies in her social class, Bower 'did the season' and spent a year going to lunches, teas, dinners and dances. She taught herself archaeology by reading books and was delighted when, aged twenty-three, a female friend, housekeeping for her brother in north-east India, invited her to visit. Her mother hoped Ursula would meet a husband on her travels but instead she fell in love with scantily dressed head-hunters – tough warriors belonging to Naga tribes. Living in remote hilltop villages along the densely forested Indian–Burmese border, the Nagas had little contact with the outside world until the British colonised their territory in the nineteenth century. In her memoirs Bower described them as 'tall, solid, muscular, Mongolian, bare chested, wearing black kilts'. The British banned their practice of beheading their enemies. Bower found it still going on but from the moment she arrived in Nagaland, she felt at home. 'This is my place. I belong here. It was as though I had rediscovered a world to which I had belonged the whole time; from which, by some accident, I had been estranged.'

When Bower announced she would hike into remote Naga villages to help dispense medicine to the sick, the local British community was horrified. 'You won't be able to stand the marching,' they warned. 'You'll get so sick that they'll have to carry you back.' She disregarded their advice and visited the Nagas regularly, taking photographs, studying the tribes' customs and lifestyles. The Nagas adored this tall and statuesque Englishwoman. In fact some believed she was a goddess. In one village, she was besieged by admirers: 'Half the population appeared to go stark, raving mad, swarmed on me like bees . . . No matter what I was doing, sleeping, eating, resting or even bathing, the hut was invaded by somebody crying. "'You are our mother, you are a goddess, there is none greater, there is none better than you."'

As the Second World War moved closer to India, British officials were quick to take full advantage of the Nagas' devotion to Bower. One commented: 'If they must have a goddess they might as well have a government one.'

The war in the Far East was not going well for the Allies. Japan had sided with Germany and, after the defiant attack on the US navy in Pearl Harbor in December 1941, Japanese forces moved into Malaya and Singapore. By April 1942 Japan controlled Burma and was expected to attack India. The fall of India would be disastrous, not just for the British who ruled the country but for the entire Allied war effort. Supply routes to China would be cut, ending that country's support of the Allies, while nationalists in India might be prompted to rise up against British rule.

In April 1942 Ursula Graham Bower found herself in a remote railway junction in Lumding, north-east India, near the Burmese border where she and a team of local Nagas ran a canteen for the unending stream of Allied soldiers and refugees passing through by train. Bower and the Nagas carried water, serviced the steam locomotives and nursed retreating Allied soldiers. 'Each morning the refugee-trains rolled in loaded to the roof,' she wrote in her memoir. 'Let no one suppose that dealing with thousands of uprooted and demoralised human beings is a kind of Church Tea. It is a dirty sweaty, frantic, navvy's job.'

She was full of admiration for the soldiers she helped:

> They were amazing. Tired, thirsty, in rags, some with reeking wounds packed into filthy compartments, they grinned and said 'Please' and 'Thank you' and 'Take it easy Miss' and carriage after carriage went out with windows solid with waving men: 'See you on the way back, Miss! Thank you for the tea. See you on the Road to Mandalay.'

This was just the start of Bower's adventures on the critical Indian front line. Her expert management of the team of Nagas at Lumding railway junction came to the attention of military officials and to her astonishment Bower found herself recruited to lead a unit of the British army's guerrilla V Force.

The Japanese were moving ever closer to the Indian border and in an effort to secure early warning of enemy troop movements, the British had set up V Force, a clandestine, underground operation. Local tea-planters, old soldiers, Indian troops and hill tribes acted as guides and spies along the 800-mile Indian–Burmese border. Bower was ordered to recruit Nagas to work with her but it wasn't always easy. They may have revered her but many distrusted the British rulers of India. In Hangrum, a village on the route of the expected Japanese invasion, she was greeted with shouts of 'Why should we fight for the British? It's a trap. They want to take us away.'

When she finally persuaded Hangrum's chiefs to support her cause, no-nonsense Bower rejected the first conscripts she was offered. 'Out of ten candidates, only three had good eyes. None was under forty. Most of them had no teeth; and one was crippled. Drawing a breath, I spoke my mind.' The next day new, more acceptable fighters appeared.

As she travelled across the steep hills of Naga country, signing up more and more men for her V Force, there was one unexpected benefit. A combination of extreme walking and lack of money to buy food caused Bower's weight to drop from 180 pounds to 150. 'I got quite slim,' she wrote to her mother. When she visited friends at Christmas she appeared so emaciated that they wrote to V Force headquarters in Calcutta demanding she receive a pay rise. Bower never dared to complain because she knew that it was unprecedented for a woman to be in charge of a British army guerrilla operation. She was always terrified that she would be fired.

In January 1943 Bower's commanding officer paid a visit. He found her organisation most impressive and was very sympathetic. He told her it was tough being female because 'you are always wrong twice – once for being wrong, and once for being a woman.' After his visit Bower's salary and her operational budget went up.

The invasion of India by the 84,000-strong Japanese Fifteenth Army finally came in March 1944. Bower's force of 150 Nagas was active in every village of the huge area she commanded. She set up a system called Watch and Ward. Runners were stationed along main routes so news and orders could be passed on quickly. Scouts watched for advancing Japanese soldiers and took anyone without a specially issued pass to headquarters. In one case of mistaken identity, a Welsh signaller was picked up as a Japanese spy because he had black hair. The Nagas had never seen a European who wasn't blond or brunette. Under Bower's command Nagas also reported crashed aircraft, and escorted any surviving Allied aircrew and soldiers through the jungle to safety.

The Japanese soon learned of her activities and put a price on her head: 'We lived like gazelles with lions about, ready to leave at once with the utmost speed,' she wrote. 'If we went to pick tomatoes, we took our tommy guns.' At night she slept with her Sten gun in shelters dug into the ground. She and her Naga volunteers buried food caches in case they were forced to hide out for prolonged periods. 'There were cattle-trails, pig-paths, game-runs, our own tunnels – every shape and condition of bolt-hole – so the chances of trapping us were few.'

Although her men were issued with ancient muzzle-loaders which Bower described as 'Last-of-Mohican guns', they were ordered not to engage with the Japanese. Their main task was intelligence-gathering, vitally important in the dense, jungle terrain which could not be monitored from the air. A typical patrol would walk along paths made by animals where Bower said, 'you couldn't see 3 feet on either side of you. Frightened the life out of me.'

A native scout would lead the way. There would be a fifty-yard gap and then the first half of the patrol, another fifty-yard gap and then the rear half usually commanded by Bower. If the group came across anything unusual, they would dive into the bushes to avoid being seen. Although they were told not to fire on the Japanese, it was sometimes unavoidable. In an interview recorded in 1985 Bower described how a brave scout, fifty yards in front of his patrol, turned a corner on a winding trail and came face to face with Japanese troops. He knew the enemy soldiers would chase him if he doubled back, so:

> He raised his muzzle loader and simply shot the nearest Jap. They thought he was a passing Naga and they opened up on him with automatic fire and of course riddled him as he knew would happen. But the automatic fire alerted his patrol who dived into ambush. The Japs walked on round the bend and his men got them. If ever a man gave his life for his friends, he did.

Another Naga stood his ground when the rest of his village wanted to retreat from the Japanese. He said it was dishonourable, that he was a subject of the king and wanted to fight for the British. Since he didn't have a gun, he took up a spear 'and went single handed for a Jap patrol and killed five of them before they shot him. He didn't mind. He was going to do his bit for the honour of the King and that was it.'

As more and more Japanese troops poured across the border from Burma, Bower needed additional firepower. She sent an urgent message demanding more weapons. When it arrived at V Force headquarters, her commanding officer:

> clapped his bush hat on his head, looked at my signal . . . and rushed around to see Slim [General Bill Slim was commander of the British Fourteenth Army, a multinational force that included British, Indian and African soldiers]. He slapped the telegram on the desk and said 'From Miss Bower Sir'. Slim looked at it and said, 'We must support her.' And I got back from a patrol to find boxes of rifles lying on the floor of my hut, and boxes of grenades, large boxes of ammunition and 150 pounds of gun powder for my men's muzzle loaders. Well I never thought that a box of grenades could look beautiful but it did. It was one of the nicest sights I've ever seen. We were up to that point armed with one shot gun and a muzzle loader.

Bower got the chance to thank General Slim in person for her 'beautiful grenades' on a visit to V Force HQ. Her thoughtful bosses encouraged her to travel to Calcutta supposedly to collect supplies but for her it was really 'an excuse for me to get a permwave and some new clothes'. On this occasion she was ordered to put on her best clothes because General Slim wanted to meet her. She thought, 'Oh heavens. My god. Here we go. Typing pool in Delhi. I'm going to be chucked out. This is the finish. Oh dear oh dear. Help.'

When she was ushered in to the general's office, she was so scared that she closed her eyes and stumbled towards Slim's desk. When she opened them 'there behind the desk, looking equally as astounded as I was, was Bill Slim. He leapt up, held out his hand, shook mine warmly and said, "Oh thank God. I thought you'd be a lady missionary with creaking stays." After which, we got on like a house on fire.'

At their meeting Slim told Bower how valuable the intelligence gathered by the Nagas had been. They discussed two brave men who had boldly walked into Japanese headquarters and asked for jobs. They were hired as water carriers but went straight to a safe containing the Japanese battle plan for Kohima, a key Naga town the Japanese were determined to take. They broke into the safe, stole the papers and handed them over to the British. The Japanese believed the Nagas were ignorant locals who couldn't read so they stuck with their plan. Slim told Bower the stolen documents had helped him turn the tide against the Japanese in the Battle of Kohima.

All through the four-month Japanese invasion, Bower's Nagas stuck by her. At one point, when it looked as if her area could be overrun by enemy troops, members of her team asked for twenty-four hours' leave to go back to their villages. 'I thought "Oh well this is it. Here they go and I can't blame them. They were never recruited for this."'

Twenty-four hours later they were all back again but without the valuable necklaces they usually wore. It turned out they'd gone home, made their wills and left their best necklaces with their sons. They returned wearing simple beads they wanted to be buried with. When Bower asked why they had done this, one Naga replied that they refused to run away and leave her to be killed. He said the Nagas could not bear the thought of surviving her death. When Bower stared at him in disbelief, he asked which was better: 'hearing people shout at our children, "Son of a coward!"? or coming back to die with you?' 'He left me speechless,' Bower recalled. 'The whole of the Nagas were completely loyal to the British. 300,000 Nagas and only two turned traitor.'

After the Japanese were defeated, they retreated back to Burma along what became known as the 'road of bones'. Nearly half of the Japanese force of 84,000 that invaded India were killed or starved to death.

With the Japanese no longer a threat, Bower's force was disbanded and she was put in charge of a jungle training camp. She and her Nagas taught Royal Air Force officers how to survive in the jungle. One pilot described her as having 'a cultured accent – and she was always very feminine. Even in the jungle, her hair was swept back and she had this clear skin . . . she was so laid back.' Another trainee said, 'We were captivated. Every one of us said later that if she said "I want you to hang yourself by the neck from the nearest tree", I am sure we would have done it. I would have followed her into the jaws of hell. She was an exceptional person.'

From the moment Bower was recruited by the British army, there had been wild rumours about the lady in the jungle. Bower remembered hearing that 'my Naga bodyguard took the head of anyone who looked at me.' In July 1945 Colonel Fredrick Betts, a fellow V Force officer, decided to see the so-called Naga Queen for himself. He visited Bower on the pretence of looking for butterflies and within four days proposed marriage. Their wedding was followed by a joyous Naga ceremony which began with the village priest strangling a rooster and taking omens from the way it crossed its feet. 'We danced. We sang till the sun went down. We danced the moon down, too,' wrote the happy bride. The omens must have been good because the Betts had a very happy life together. They remained in India until after the war and then moved to Kenya where they grew coffee. In 1952 they moved back to Britain, settling on the Isle of Mull with their two daughters.

In his memoirs, General Slim paid tribute to the Nagas' wartime effort, a tribute that applies equally to the woman who led them:

> They guided us, collected information, ambushed enemy patrols, carried our supplies, and brought in our wounded under the heaviest fire, and then, being the gentlemen they were, often refused payment. Many a British and Indian soldier owes his life to the naked head-hunting Naga, and no soldier of the Fourteenth Army who knew them will ever think of them but with admiration and affection.

Ursula Graham Bower received the Lawrence Memorial Medal, named after Lawrence of Arabia, for her anthropological work among the Nagas. Her diary, letters and some of the 1,000 photographs she took in Nagaland can be seen at the Cambridge University Museum and London's Horniman Museum. Among the artefacts she donated to the Pitt Rivers Museum in Oxford is a Naga head-hunter's shield complete with human hair.

COUPLES

FOR OBVIOUS REASONS we rather approve of couples who do a lot together. Tales of women going to war with their husbands or being left behind at home range from the tragic to the triumphant. When Semiramis, Queen of the Assyrian Empire in the ninth century BC, lost her husband Ninus in battle, she took over command disguised as her son, the king's successor. Fooled by her impersonation but inspired by her leadership, her army routed the enemy. During Napoleon's Austerlitz campaign in 1805, Madeleine Kintelberger was beside her husband, a soldier in the French army, when he was killed by a Russian cannon ball. Six months pregnant, she picked up his sword and ran at the advancing Cossacks. They shot her in the leg but she charged on. A second shot in her other leg finally brought her down. She survived and gave birth to twins in a Russian prison camp.

For every wife who had the opportunity to fight, there were thousands more who slogged along loyally after their fighting men. When we visited the battlefield of Culloden near Inverness it was painful to see a patch of grass where, we were told, the wives and families of Scottish soldiers sat and watched the bloody battle in 1746. It took less than one hour for the troops of King George II to annihilate the men loyal to Bonnie Prince Charlie and then kill most of their women and children too. During the American War of Independence General George Washington wrote: 'the multitude of women, especially those who are pregnant, or have children, are a clog upon every movement.'

By the nineteenth century the numbers of women who accompanied their husbands on military campaigns were reasonably well regulated. When the Duke of Wellington was sent to defeat Napoleon's forces occupying Portugal and Spain in 1808, a handful of wives per battalion were chosen by lot to travel with their husbands. The women worked as cooks, laundresses and nurses. Each was given half a man's rations, a child received one quarter. We've chosen to tell the story of one Spanish girl of fourteen who met Captain Harry Smith at Wellington's siege of Badajoz: she married him and followed him all the way to the Battle of Waterloo.

As well as the captivating story of Harry and Juana Smith, we've selected two other remarkable accounts of couples who've endured the horrors of war. Helen Thomas described in heartbreaking detail how it felt to say goodbye to her beloved husband as he left to fight in France in 1917. And Florentia Sale, wife of a British major-general, told the dreadful story of how she survived being prisoner of an Afghan warlord during the First Afghan War in 1841–2.

13

Sir Robert and Lady Sale

Terror in Afghanistan, 1841–2

ON 13 JANUARY 1842 one of the few British survivors of an army of some 16,000 men arrived alone, wounded and utterly exhausted at the gates of Jalalabad in Afghanistan. William Brydon, who'd been the army's doctor, almost fell from his saddle as people ran to help him. He revealed to them the sorry story of one of the greatest military disasters in British history, the annihilation of an army in the First Afghan War. It was the first in a series of alarm bells for a foreign imperial power intent on controlling this vast intransigent country in central Asia.

The story begins three years earlier when, in pursuit of this strategic dream, the British occupied Afghanistan and sent its ruler into exile. The Afghans riposted with a popular uprising led by the ruler's implacable son, Akbar Khan. He and his followers laid siege to the British in Kabul, forcing them, their families and servants to abandon the capital, and head east towards India. Men, women and children embarked on what was to become a suicidal attempt to escape.

Dr Brydon told how in the course of the attempted march to safety all the British men except him were killed in a series of bloody clashes. But where were the women and children? All Brydon knew was that they had been promised protection by the insurgents early in the flight and had been taken away as captives. He had no idea what had happened to them. The most senior and, as it turned out, formidable of the women was Florentia Sale, the wife of Major-General Sir Robert Sale. He was a fine soldier, whose habit of always leading his men resolutely from the front had earned him the nickname 'Fighting Bob', but he was not with her when the army started its ill-fated march. He'd been sent ahead to secure a route for the retreat, leaving his wife and nineteen-year-old daughter Alexandrina behind in Kabul.

Sale was anxious to rescue them but soon found himself penned inside the fortress of Jalalabad, ninety miles from Kabul on the road to India. He wasn't to see his loved ones for another eleven months – months in which his wife and daughter were to endure untold torment and horror. Often accounts of such hardship are lost but we have the great good fortune that it is all recorded in Florentia Sale's extraordinary journal. She writes in the manner of the hard-nosed modern war correspondents we have both worked with in our careers as journalists. She gives a no-frills account of the hardships and indignities she, Alexandrina and the other women suffered. She spares no one in her description of the cruelty and deceit of her captors and their leader Akbar Khan or the blockheaded stupidity of many of the British officers. She is appalled at the blundering and arrogance of a government that landed its forces in such a catastrophic predicament. There have been many victims of Britain's wars in Afghanistan right up to the present. Lady Sale had to endure the first of the fruitless Afghan wars. And she was determined that the world would know about it.

Florentia had married Robert Sale in 1809 when she was nineteen. The daughter of an East India Company civil servant at a time when the company effectively ruled India, she was already a formidable character and more than a match for Fighting Bob. Her family called her the 'Grenadier in Petticoats'. She and Robert settled happily into married life and had twelve children. Half of them died young, as happened so often in those days. By the late 1830s many of the others had left home, but Alexandrina still lived with her parents, enjoying the pampered life of a well-to-do military family in British India.

Then suddenly the Sales found themselves thrust into a key role in a great British imperial adventure. Under the influence of Foreign Secretary Lord Palmerston, who was determined to pre-empt any Russian ambitions to expand further into central Asia and threaten British India, the government launched an occupation of Afghanistan in March 1839. The country was then as now a hotbed of quarrelling tribes united loosely under the ruler in the capital Kabul. Robert Sale, by now a major-general, was sent to Kabul as second-in-command. Like many officers at the time he was accompanied by his family and they expected to continue to enjoy all the luxuries of their lives in British India. They made the trip with scores of servants and animals. The Sales even took their grand piano. Life in Kabul was as pleasurable as in Bombay or Calcutta. There was hunting, sightseeing, cricket, parties and even gardening. Lady Sale writes in her diary: 'My sweet peas and geraniums were much admired.' Her daughter Alexandrina, although described as 'ignorant and illiterate' by one old cynic, was the only unmarried young woman around. She was much admired by young officers and soon married one of them, Captain John Sturt.

Fighting Bob's boss was General William Elphinstone, who suffered from gout and whose incompetence was to become legendary. Britain's political envoy in Kabul was the haughty Sir William Macnaghten and Alexander Burnes was the political resident. Their job was to protect Shah Shuja, the ruler Britain had foolishly imposed on the country. He commanded little respect from most Afghans who resented the fact that their former ruler, Dost Mohammad, had been forced out of power. It was blatant regime change and it was to end in disaster. Lady Sale was to have a unique vantage point from which to witness the turbulence. She believed Elphinstone should be sacked for being so feeble and thought the two British political leaders in Kabul hopelessly defeatist. 'There is much reprehensible croaking going on,' she writes.

By 1841 the country was on the edge of insurrection and the British were sitting on a powder keg. In Kabul itself Shah Shuja was holed up in its fortress, the Bala Hissar, which can still be seen today, and the British forces lived in scattered cantonments below. As attacks by Akbar Khan's insurgents turned into a siege, the decision was taken to evacuate Kabul. General Sale was sent off in October 1841 to clear the route for the rest of the force, 'fully expecting me to follow,' writes Florentia Sale in her journal, 'in three days'. Robert Sale's force had a nightmare journey. It was assailed by Afghans and suffered heavy losses: he himself had his leg shattered in one ambush. He coolly told one of his aides, 'I've got it', but he remained on horseback until loss of blood made him hand over command. He and what was left of his brigade eventually made it to Jalalabad fifty miles away but, surrounded by enemy forces, they were trapped and unable to come to the rescue of those they'd left behind.

Back in Kabul, Florentia Sale quickly saw that her husband's promise to return in three days would not be realised any time soon. Attacks on the British community increased day by day. On 2 November she was shocked to see her daughter's husband John Sturt carried into her house with deep stab wounds in his shoulder and side, and even on his face. His throat was choked with blood, and Florentia and Alexandrina managed with great difficulty to get a spoonful of water into his mouth 'by a drop or two at a time, painful as it was for him to swallow it'. At the same time the political resident Sir Alexander Burnes disappeared and his dismembered body was later found strung up on the trees in his garden. On Christmas Eve Lady Sale had the distressing job of breaking the news to Lady Macnaghten that her husband, the British envoy, had been murdered at a meeting with Akbar Khan, whom she describes in her journal as 'proverbially the most treacherous of men'. It appears that Akbar took hold of Macnaghten, grappled with him and threw him on the ground. Akbar fired his pistol at him and wounded him, and the envoy was immediately cut to pieces. Later Akbar claimed he had no hand in Macnaghten's death. 'To prove his sincerity,' Lady Sale writes sarcastically, 'he wept for two hours.' The body was paraded around the city and displayed in the Great Bazaar.

By the end of December bloody raids on the British camp and the obvious imbalance of forces that made the enemy more and more menacing prompted General Elphinstone to agree a treaty with Akbar which Lady Sale calls 'most disgraceful'. The Afghan leader was to be paid 40,000 rupees and the army would surrender all but six of its guns and allow six officers to become hostages. In return Akbar undertook to protect the army and its followers from attack, to provide it with food and fodder and to escort it from Afghanistan. The deal would have been shameful enough if it had been respected by Akbar. But it was not.

On 6 January 1842, a clear and frosty day with a foot of snow on the ground and the temperature below freezing, a small group of Afghans witnessed the sorry spectacle of a British army of 4,700 men and 12,000 followers, including the women and children, snaking out of Kabul. They were headed for the safety of Jalalabad on the road to the Khyber Pass and India. The first blow was the loss of nearly all the convoy's baggage and supplies as the servants, frightened by plundering Afghans, threw away their loads and fled. At the end of day one all that the two Sale women had left was Alexandrina's bedding. The column halted at 4 p.m. after making little progress. There were no tents and it was intensely cold. Everyone scraped away the snow as best they could to make a place to lie down. Lady Sale and her daughter were lucky enough to have an officer pitch a makeshift tent over them. 'But it was dark and we had few pegs; the wind blew in under the sides, and I felt myself gradually stiffening . . . I felt very grateful for a tumbler of sherry, which at any other time would have made me feel most unladylike.' Within twenty-four hours 650 men had deserted. Several more were found frozen to death at dawn. The retreat had got off to such a dreadful start that Florentia Sale reckoned she, her daughter and the other families had only a meagre chance of survival.

Progress was no better the next day. 'Numbers of unfortunates have dropped, benumbed with cold, to be massacred by the enemy.' There was worse to come. The Afghans were now firing at the shrinking force from all directions and from just fifty yards away. Alexandrina's pony was wounded in its ear and neck. As for Florentia Sale herself, 'I had only one ball in my arm; three others passed through my poshteen [sheepskin] near the shoulder without doing me any injury.' They only escaped more serious injury by pushing their horses as fast as they could go. Other women further back came under heavy fire. Mothers and babies were separated. Many of the women who survived the shooting were carried off or disappeared. Fifty regular troops and 2,500 camp followers were killed that day. Alexandrina's husband James Sturt, who'd recovered from his earlier wound, suffered another injury which this time was to prove fatal. 'Poor Sturt was laid on the side of a bank with his wife and myself beside him.' A doctor tended to her son-in-law's wound: 'He dressed it; but I saw by the expression of his countenance that there was no hope. He afterwards kindly cut the ball out of my wrist, and dressed both my wounds.'

The following day, 9 January, Sturt's discomfort was clearly intolerable even when they found a pony to carry him. 'The rough motion increased his suffering and accelerated his death; but he was still conscious that his wife and I were with him, and we had the sorrowful satisfaction of giving him Christian burial.' By this time nearly half the surviving force was frostbitten or wounded: everyone was bitterly cold and hungry.

That morning Akbar Khan made a proposal. He offered to take all the women and children under his protection and escort them to Peshawar in India. Many of the officers, including the commanding General Elphinstone, argued the offer should be accepted. Overcome with grief for James Sturt, neither Florentia nor Alexandrina Sale were in a fit state to decide for themselves whether to agree to this proposition from a man they deeply distrusted. But they felt they had no choice. 'Our present position is one of imminent peril . . . There was but a faint hope of our ever getting safe to Jalalabad; and we followed the stream.' From then on the women, children and a handful of British officers were separated from the rest of the army.

When the Sales reached a fort where Akbar provided them with three small dark and dirty rooms, they found themselves in the company of six other women and nine children. 'At midnight some mutton bones and greasy rice were brought to us. All that Mrs Sturt and I possess are the clothes on our backs when we quitted Kabul.' They walked off the next day under escort and immediately found themselves passing grim scenes where the main force they'd just been with had suffered dreadful slaughter. 'The road [was] covered with awfully mangled bodies, all naked: fifty-eight Europeans were counted . . . the natives innumerable . . . numbers of camp followers, still alive, frost bitten and starving; some perfectly out of their senses and idiotic.' That night the Sales and their companions were crammed into a little room in another fort. 'An old woman cooked chupattis for us, three for a rupee; but finding the demand great she soon raised the price to a rupee each.'

As well as describing her own sufferings under the 'protection' of Akbar Khan, Lady Sale recounts the tragic fate of the main British force which had left Kabul at the same time and was trying separately to fight its way through to Jalalabad. By 10 January she says 12,000 of them were dead and only around 4,000 still alive. Akbar offered the army commander General Elphinstone safe passage if he agreed to give up all remaining weapons. He refused. The following day the unfortunate Elphinstone and two of his officers were invited to meet Akbar in his bivouac. They were greeted with great courtesy and treated to a slap-up meal by the Afghan leader who explained that he was unable to contain the fury of the local tribesmen who had an intense hatred of the British. The general and his two companions were told they would be held as hostages. Akbar promised them safe passage provided that Lady Sale's husband, Major-General Sir Robert, agreed to evacuate Jalalabad. Elphinstone again refused.

Meanwhile the attacks on the fast dwindling army increased. The soldiers struggling to fight off the Afghan tribesmen had little to eat or drink. All they had was a scanty meal of camel's flesh. The horses were starving: some were reduced to eating each other's tails. There were two further major attacks on the army on 11 and 12 January and by the 13th it had shrunk to twenty officers and sixty men. A final stand was made on a small hill. Several times the Afghans attempted to dislodge them and were repulsed. But then the ammunition ran out when there were only a handful of them left. 'The enemy made a rush,' writes Lady Sale, 'which in our weak state we were unable to cope with. They bore our men down knife in hand.' A little later that day Dr Brydon arrived alone at Jalalabad.

Meanwhile Lady Sale, her daughter and the other women and children still faced eight months of captivity in the hands of the duplicitous Akbar Khan. By 17 January, eleven days after leaving Kabul, the awful truth was dawning on Florentia Sale that he had no intention of taking them to safety. They were now virtually prisoners and she lost all hope of being reunited with her husband Robert who was still holed up in Jalalabad. Her only concession was that she was to be allowed to write to him, receive his letters and a parcel of clothes. Robert was still in effective control of Jalalabad and had been reinforced. He had fought off Akbar's attempts to storm the city but his force was not strong enough to stage a rescue mission for his wife and the other families. One of his staff said later that what Sale feared more than anything else was that Akbar Khan would drag his wife before the walls and torture her within his sight. 'If that happens,' said Sale miserably, 'I will have every gun turned on her . . . I will never surrender.'

In the middle of February, and no closer to freedom, Florentia and her daughter were caught up in an earthquake. Lady Sale was being held captive on the top floor of a house when 'the roof of our room fell in with a dreadful crash. The roof of the stairs fell in as I descended them; but did me no injury.' She was immediately desperately anxious about Alexandrina being buried: 'I could only see a heap of rubbish . . . I was nearly bewildered,' but she then heard the cry, 'All are safe.'

Her contempt for Akbar Khan, her captor, was redoubled when she heard what he had done to one of his close friends whose firearm had gone off by mistake as he was dismounting, wounding Akbar in the arms and lungs. Lady Sale writes that the unfortunate friend was instantly cut to pieces or according to another account 'burned alive . . . There is nothing too brutal or savage for Akbar to accomplish.'

Three months after leaving Kabul, Lady Sale learned that Shah Shuja, the ineffectual ruler the British imposed on Afghanistan, had been assassinated. By the beginning of April 1842 Akbar was still keeping the British women on the move under his 'protection'. Soon after, Lady Sale was attacked by a fever which lasted for several days. 'I was utterly incapable of sitting on horseback . . . My turban and habit were completely saturated by the rain; and I shivered as I went.' On 24 April she discovered that General Elphinstone had died in the night, the climax of a long illness that may partly explain his failure of leadership. His body was sent by Akbar to Jalalabad for burial but the porters were attacked and the coffin, which the assailants thought contained treasure, was broken open. 'We at first feared they had mutilated the poor old man's body; but only a few stones were thrown, one of which struck the head.' What was left of

Elphinstone was eventually buried with full military honours in Jalalabad.

In spite of all the hardship and the daily threats of a violent attack, Florentia Sale remained defiant. 'I am not at all inclined to have my throat cut. On the contrary,' she writes, 'I hope to see the British flag once more triumphant in Afghanistan.' She had no objection to Akbar's father, Dost Mohammad Khan, being reinstated. 'Only let us first show them that we can conquer them, and humble their treacherous chiefs in the dust.'

At the end of May, Lady Sale and her daughter found themselves back where they'd started – in Kabul – where tribal infighting was taking place and rumours abounded that the British were about to send an army to retake the capital. Spring turned to summer and on 24 July Florentia Sale notes briefly: 'At two p.m. Mrs Sturt presented me with a grand-daughter – another female captive.' At last the news broke that two British columns were advancing to retake Kabul. Lady Sale agonised about what Akbar might do to her and her daughter. 'What will now be our fate seems very uncertain; but I still think he will not cut our throats – not out of love to us, but because the other chiefs would resent it, as, having possession of us, they could at least obtain a handsome sum as our ransom.'

By now Akbar had the women back on the road again and Lady Sale was determined that, if she survived, she would publish her journal. She urged her future readers – with words that have a familiar ring to us today nearly two centuries later – to recognise that the best solution in Afghanistan was to allow the Afghans to run their own country with the ruler they desired. In her journal she indicates her contempt for a policy of 'attempting to keep possession of a country of uncivilised people, so far from our own'. But she rejects the idea of admitting defeat. She calls for revenge for 'the foul murder of our troops . . . do not let us dishonour the British name by sneaking out of the country like whipped Pariah dogs.' By now she was confident that she would be rescued. Her joy was unalloyed when she heard that Kabul had been reoccupied and that her husband was on his way with a brigade to rescue her. She says she couldn't find words to express her feelings on her husband's approach. 'To my daughter and myself happiness so long delayed as to be almost unexpected was actually painful, and accompanied by a choking sensation, which could not obtain the relief of tears.' When she was reunited with him, she was congratulated by each of the soldiers of Sale's Light Infantry, the 13th Regiment of Foot (later the Somerset Light Infantry, whose cap badge carried the emblem 'Jellalabad') she could no longer contain her tears that 'now found their course'. She had not seen her husband for eleven months.

With the return to India in December 1842 of the so-called 'Army of Retribution' that recaptured Kabul, Britain felt able to make the best of failure and end the First Afghan War. Dost Mohammad was restored to power in Kabul. Lady Sale and her husband, Fighting Bob, were about the only two Britons to emerge from the conflict as heroes. Florentia Sale's journal had now reached London and was serialised in *The Times* and published as a book. When the Sales arrived back in Britain they were feted in *The Times* of 26 July 1844 as 'Major-General Sir Robert Sale, the equally heroic Lady Sale and their widowed daughter, Mrs Sturt and child'.

Florentia Sale's account, widely published in Britain and Europe, did much to highlight the mistakes of the First Afghan War and the dangers for the British Empire of plunging into imbroglios that had no easy solution. But the task of keeping Afghanistan out of the hands of an unfriendly power – an obsession that became labelled the Great Game – long outlived any advice from people like Lady Sale and persists even until today. The lessons of the disaster she witnessed and described so well in 1842 have still not been fully learned. The Sales were back in India at the end of 1844 and Sir Robert was mortally wounded in the First Sikh War a year later at the age of sixty-three. Florentia stayed on in Simla and saw her daughter married again to another army officer. But fate caught up with Alexandrina even more ferociously than before: she and her husband were victims of the Indian Mutiny in 1857. As they sat in their carriage in Bihar they were attacked by four mutineers and beheaded. It was a tragedy that Florentia Sale was spared from experiencing. She had died on a visit to Cape Town in 1853 and was buried in a simple tomb with a short inscription: 'Underneath this stone reposes all that could die of Lady Sale'.

14

Harry and Juana Smith

From Badajoz to Waterloo, 1812–15

ANOTHER BRITISH MILITARY couple with a highly adventurous past happened to be in India at the same time as the Sales were in Afghanistan. Colonel Harry Smith, whose wife Juana was Spanish, was aghast at the news from Jalalabad in 1842 and lobbied hard with the British authorities in India to allow him to lead a rescue bid for his old friend General Elphinstone. But the high command in India was inclined to be cautious, as Smith himself recounted: 'When the most vigorous . . . steps ought to have been taken with the velocity of lightning . . . the moment was lost . . . it was perfectly practicable, as I then pointed out.' Harry Smith was nothing if not sure of himself. He was asking the British commander-in-chief in India to give him full military command and authority in Afghanistan, when he had no previous experience of the place. His presumptuous bid was rejected.

Harry Smith and his exceptional wife, Juana, had enjoyed more than their share of excitement. They had met thirty years earlier in extraordinary circumstances. Juana María de los Dolores de León was a fourteen-year-old member of a distinguished Spanish family living in the fortress town of Badajoz near the Portuguese frontier. Its massive fortifications made it a natural stronghold for the French Emperor Napoleon's expanding empire. His forces seized it when his army occupied Spain and Portugal in 1807. Britain was locked in a struggle for survival with Napoleon and in 1808 Sir Arthur Wellesley – later the Duke of Wellington – landed an army in Portugal intent on liberating Spain and Portugal and pushing the French back behind their border. It was the beginning of what became known as the Peninsular War.

Nearly every step Wellesley took brought success – and Harry Smith closer to meeting the love of his life. A handsome, engaging, twenty-one-year-old with a bit of a swagger, he was not tall; he was neat and elegant with curly hair, an accomplished horse rider who loved the hunt. By the time Wellington's forces moved into Spain Smith was a lieutenant in the elite 95th Rifles: they were sharpshooters trained to fire the new rifle which had a range far in excess of the ordinary infantryman's musket. He had already served in South America so spoke fluent Spanish which served him and Wellesley's army well as it struggled to communicate with Spanish guerrillas resisting the French occupation.

By 1812 Smith, aged twenty-four, was a captain and a seasoned company commander. Wellesley, now honoured with the title Lord Wellington, had pushed the French out of Portugal and was advancing east into Spain. The fortress of Badajoz stood in their way, powerfully placed on the River Guadiana commanding a major road to Madrid. It had to be suppressed and the 95th would be in the forefront of Wellington's assault troops. Smith was in the thick of dreadful butchery as his regiment tried to storm through a breach. Badajoz's high walls were well defended, even when they were battered by Wellington's cannon, and the French exacted a terrible toll. Wave after wave of British infantrymen were cut down and Wellington and his staff were near despair. 'It was appalling. Heaps on heaps of slain – in one spot lay nine officers,' wrote Smith. Then in the early hours of 7 April General Thomas Picton's 3rd Division scaled the walls and the French defence collapsed. Badajoz fell and the road to Madrid was open.

But the price paid by Badajoz's garrison and its innocent Spanish civilian population was horrific. Harry Smith witnessed enraged British soldiers brutally avenging their fallen comrades. 'The atrocities committed by our soldiers on the poor innocent and defenceless inhabitants of the city no words suffice to depict,' he wrote. They raped, pillaged and drank. Harry Smith and a friend were out in the streets of the ravaged city when two young Spanish women came towards them with blood streaming from their ears. The two officers, horrified at the women's condition, asked what had happened. 'Your rampaging soldiers tore off our earrings,' they replied. At which Smith and his friend immediately insisted on taking the pair under their care. The younger woman was, in the words of Harry Smith's friend, 'transcendingly lovely . . . with a face so irresistibly attractive . . . that to look at her was to love her'. Smith obviously agreed because no sooner had they escorted the women back to the camp than he proposed to fourteen-year-old Juana and she promptly accepted the man she called Enrique. Most of his fellow riflemen thought he'd taken leave of his senses. She was a Catholic, educated in a convent school, he was a Protestant, a professional soldier approaching the peak of a very active career: the last thing they thought he should do was saddle himself with a fragile young wife accompanying the army through the peninsula.

But Juana Smith turned out to be anything but fragile. She threw herself into the hardships of camp life and long-distance marching as if she'd spent her short life preparing for it. The Duke of Wellington put his official stamp on the marriage only twelve days after the couple met by agreeing to give Juana away at the wedding ceremony.

She was utterly devoted to Smith, as he was to her, and she soon won the admiration and affection of his comrades. In his autobiography written nearly half a century later he described her as his 'guardian angel'. For the next two years, during the searing summers and bitter winters of the peninsula and the Pyrenees, Juana followed Harry through the rigours of the campaign that eventually threw Napoleon's armies out of Spain.

Harry's first task was to find his new wife a mount. She had only ever ridden a donkey. Offered a placid Portuguese horse from Smith's hunting stable, she preferred a large Spanish thoroughbred called Tiny and was soon as agile in the saddle as a cavalry trooper. After each battle Juana, like other army wives, had the distressing job of searching the field for her husband. At the Battle of Vitoria in the summer of 1813, Harry's horse collapsed on top of him and the rumour went around that he was dead. Juana was desperately anxious and it was some time before she discovered that neither Harry nor the horse was hurt. He'd scrambled out from under the horse and revived it with a kick on its nose.

Juana's stamina and charm throughout this exacting hardship delighted Harry and captivated the officers and men he campaigned with, including Wellington himself. 'There was not a man who would not have laid down his life to defend her,' wrote Smith later.

Wellington's army crossed the Pyrenees in the winter of 1813, slowly but surely pushing Napoleon's forces out of Spain after the Battle of Vitoria. He went on to defeat the French in southern France at the River Nive in December, at Orthez in February 1814 and finally at Toulouse in April. Shattered by this relentless British advance and – more importantly – by the overwhelming weight of his opponents in central and eastern Europe who defeated him at the Battle of Leipzig in October 1813, Napoleon abdicated in April 1814 and was exiled to Elba. Europeans sighed with relief, and the victors made ready to go home. For the Smiths it should have been the end of daily anxiety with the prospect of a carefree life in England. But, wrote Smith: 'My happiness of indolence and repose was doomed to be of short duration.' His commanding officer summoned him and told him that he was off to America. It was the only campaign on which Juana would not accompany him.

The War of 1812 was a tiresome conflict that the United States started when Britain interrupted American trade with France. The Americans, furious that the Royal Navy was interfering with their ships, declared war on Britain and invaded its colony of Upper Canada. With Napoleon out of the way the order came from Whitehall: the Americans must be given 'a good drubbing'. One of Wellington's respected peninsular veterans, Major-General Robert Ross, was to command a task force that would invade the eastern seaboard of the USA, and Harry Smith was a natural choice for a senior role on Ross's staff. Smith was desperately torn: as an aspiring officer he had no choice. 'But I knew I must leave behind my young, fond and devoted wife, my heart was ready to burst, and all my visions of our mutual happiness were banished in search of the bubble reputation.' Juana was distraught but recognised that Harry could not miss such an opportunity.

The couple spent a few days drifting down the Gironde from Bordeaux to the sea in a small skiff. It was a delightful trip and the beauties of the scenery and the spring foliage did much to relieve Juana's worry about her uncertain future in a strange country. Harry's brother Tom undertook to look after her, and the Smiths parted at the end of May. 'I left her insensible and in a faint,' recalled Harry, as he sailed off across the Atlantic in the seventy-four-gun *Royal Oak*.

Juana was glad to have the company of Tom, who'd served in the peninsula, as she travelled to Britain for the first time. Tom found her accommodation in London and told her that her horse, Tiny, was happily grazing with the other Smith horses and ponies at their home in Whittlesey, Isle of Ely. Wouldn't Juana like to go there and meet the family? No, she said, she was too shy to face them until Harry came home: she was happy to stay in London and study English with a tutor. It wasn't till September – four months after she'd arrived in England – that she got her first letter from Harry.

It had been written before he and the army arrived in America. By the time the letter reached Juana the astonishing military adventure in which Harry played a major role was over. On 20 August Major-General Robert Ross landed his force of 4,500 British troops, most of them grizzled veterans of the Napoleonic Wars, on a riverbank fifty miles south-east of Washington, DC. This city was the brand-new capital of the thirty-year-old United States. It had a tiny population of only 8,500 people but boasted the two most precious buildings in the new nation: the Capitol that housed the two houses of Congress and the President's House, already being called the White House.

There was immediate pandemonium and panic in Washington as Ross marched his army briskly up to the crossing of the eastern branch of the Potomac at a small town called Bladensburg. Drawn up on the other side of the river was the 6,500-strong US army, largely composed of poorly trained militiamen. The battle that followed was one of the most shameful defeats in US history. Faced with the battle-worn redcoats the Americans fled, abandoning the field and the city of Washington to the British. Smith was one of the forty or so British officers who accompanied Ross into the empty White House abandoned by the President, James Madison, and his wife Dolley. In their rush to leave the President's staff had left dinner on the table, the meat still roasting on the spits. Harry Smith and the others sat down and made short work of it and then, piling the chairs on the table, they set light to the great mansion. 'I shall never forget the majesty of the flames,' Harry remarked, but at the same time he admitted that he personally regarded the burning of Washington as 'barbaric'. To complete the utter humiliation of Madison's government, the British went on to torch Congress, the Treasury, the State Department and the War Department.

Within days Harry was given the important mission of returning to London to report on what Ross could fairly claim to be a triumphant success. He felt doubly blessed: he would be the bearer of great news and he would see Juana again. Washington was burned on 24 August: on 22 September his wife Juana stepped out of her house into Panton Square and saw a coach draw to a stop beside her – its passenger's arm waving to her out of the window. '*Oh, Dios la mano de mi Enrique*,' she cried, 'Oh my God, my Enrique's hand.'

'Never shall I forget that shriek as we held each other in an embrace of love few can ever have known,' recalled Harry.

The next day Smith was off to take news of the victory to the Prince Regent. Harry was pleased to sense that the prince seemed to be as shocked as he himself had been that a British army had burned down the shrines of democracy in the United States. As he carefully backed out of the room as instructed by the War Minister, Harry was delighted to hear the prince say to the minister, 'Don't forget this officer's promotion.' Before he went back to join Juana that night, Smith dined at the minister's house and he remarked to his next-door neighbour at the table how well he knew and admired the Duke of Wellington. 'I am glad to hear it,' said Harry's neighbour, 'he is my brother.'

Smith had promised Major-General Ross that he would also visit Ross's wife, Elizabeth, and he and Juana set off to Bath the next day to meet her. They carried the heartening news of Ross's success in Washington: the tragedy was that even as they chatted joyfully, Elizabeth's husband was dead. As his army approached Baltimore, determined to deliver the same punishment he had inflicted on Washington, he was killed by an American sharpshooter's rifle shot. Smith claims – with hindsight in an autobiography written several years later – that he had warned Ross not to attack Baltimore. Whether or not we believe him, the enterprise was a failure. The British were rebuffed by far stronger opposition than they'd encountered at Washington. The fact that, when the bombardment was lifted, it was not the Union Jack flying over the city's fortress but America's star-spangled banner inspired the words of the US national anthem.

Smith now finally persuaded his young wife that she should meet his family and they all spent a very happy three weeks in Whittlesey with Juana's English improving by the day. But a letter from Whitehall soon parted the couple again. After the failure of Baltimore and the death of Ross, a new effort was to be made to bludgeon the Americans into agreeing an end to the war. Wellington's brother-in-law, Major-General Edward 'Ned' Pakenham, would lead a reinforced army against the city of New Orleans in order to seize and control the mouth of the great Mississippi River. Once again Harry would be a senior staff officer. 'It once more raised that blighted word "separation" to be imparted to my faithful and adoring wife.' This time Juana stayed with his family in Whittlesey.

The battle of New Orleans in January 1815 was an even greater disaster than Baltimore. The bloodshed that followed the British throwing themselves at the American lines in huge numbers has been immortalised in Lonnie Donegan's hit song, 'The Battle of New Orleans'. But the attack was badly mismanaged. The British assault on a well-fortified US line on the banks of the river was soundly beaten off and Andrew Jackson's Americans were victorious. Smith was sent bearing a flag of truce to ask the Americans to allow the British to bury their dead. 'They fired on me with cannon and musketry, which excited my choler somewhat,' but he finally won agreement from General Jackson. Smith then found himself shovelling 200 bodies into a specially dug hole in the ground. 'A more appalling spectacle cannot well be conceived than this common grave, the bodies being hurled in as fast as we could bring them.'

The sad irony about New Orleans is that two weeks before the battle was fought the British and Americans had signed a peace treaty at Ghent in Belgium. It took a month for the news to reach the USA. Some 700 men had died in vain.

In March, with the American war over, Harry Smith was on his way home again – once more aching to see his wife – but no sooner had his ship reached Britain's coast than he heard the news – shouted across the water from a vessel going the opposite way – that Napoleon had escaped from Elba. 'Such a hurrah as I set up, tossing my hat over my head! I will be a Lieutenant Colonel before the year's out,' Harry recalled exclaiming at the time. He was asked by a commander he had impressed in the American campaign to accompany him as his brigade major to the new front against Napoleon in Belgium. But this time he was determined to take Juana with him. He had bought two horses at Newmarket, one for Juana, 'a mare of great celebrity', and one for himself and they were shipped with the Smiths to Ostend. Harry reported that, far from being apprehensive at the prospect of battle, Juana was eager to accompany him. 'My wife was delighted to be once more in campaigning trim.'

Harry was with the Duke of Wellington on the morning of 18 June on the ridge of Mont St Jean just south of the village of Waterloo: 'It was delightful to see his Grace that morning on his noble horse Copenhagen – in high spirits and very animated but so cool and so clear on the issue of his orders.' Smith had told Juana to await him in Brussels a few miles to the north. Napoleon in a lightning campaign had marched from Paris with 120,000 men and crossed into Belgium intent on breaking through to Brussels. Two armies stood in his way, Marshal Blücher's Prussians and Wellington's mixed army of allies and British. In a brilliant stroke Napoleon had thrown his army against the Prussians and forced them from the field at the Battle of Ligny on 16 June. Blücher, beaten but still defiant, retreated to a town named Wavre where he judged he could still march to Wellington's aid if Napoleon attacked him. And he actually promised the duke that he would do so.

On the morning and afternoon of the 18th, believing that the Prussians were safely out of reach, Napoleon made a series of massed frontal attacks on Wellington's ridge. Smith was constantly racing to and fro carrying orders to his brigade. The battle was as fiercely fought as any in history: Wellington described it as 'hard pounding'. His task was to hold the line of the ridge against repeated French infantry and cavalry attacks until Blücher could come to his aid. 'Every moment,' wrote Smith, 'was a crisis . . . Every staff officer had two or three (and one four) horses shot under him. I had one wounded in six, another in seven places, but not seriously injured.'

Blücher arrived in the nick of time, forcing back Napoleon's right, and the French Emperor was unable to press Wellington as hard as he would have liked. The final climax, witnessed by Harry Smith from afar, came with the crushing of the French Imperial Guard's assault at around 8 p.m. Utterly defeated, the French fled from the field and Napoleon rode back to Paris and eventual exile on the island of St Helena in the South Atlantic. Blücher and Wellington shook hands and then took stock of the appalling aftermath of the battle. Some 45,000 dead and wounded lay on the field. 'I had never seen anything to compare with what I saw,' said Smith. 'At Waterloo the whole field from right to left was a mass of dead bodies.' He saw French horsemen in their metal breastplates literally piled on each other. Many unwounded men were still trapped under their horses. Others, fearfully wounded, struggled to free themselves from the horses lying on top of them. 'The sight was sickening, and I had no means or power to assist them.'

If it was a wretched experience for Harry Smith, it was torture for Juana. She had ridden all the way to Antwerp on the advice of an officer who said she would be safe there. When she heard of the British victory she raced back to Waterloo searching for her husband. Imagine her horror when she ran into a group of 95th riflemen and they told her that Brigade-Major Smith had been killed. She was beside herself with distress and rode frantically around searching for his body. She wrote later: 'I approached the awful field of Sunday's carnage, in mad search for Enrique. I saw signs of newly dug graves and then I imagined to myself: "O God he has been buried and I shall never again behold him."'

Then suddenly she ran into an old friend, Charlie Gore, and asked desperately, 'Oh where is he, where is my Enrique?'

'Why, near Bavay by this time, as well as ever he was in his life; not wounded even.'

'Oh dear, Charlie Gore, why thus deceive me? The soldiers tell me Brigade-Major Smith is killed.'

'Dear Juana, believe me; it is poor Charles Smyth,' replied Gore, 'Brigade-Major to Pack . . . Why should you doubt me?'

'Then God has heard my prayer,' said Juana, with massive relief. She didn't reach Bavay till the next morning. 'Oh Gracious God I sank into his embrace, exhausted, fatigued, happy and grateful.'

It had been the most harrowing moment in one of the most affecting of relationships: from then on the Smiths were not parted for long. Harry Smith went on to enjoy a long and successful career that took him to France, Canada, Jamaica and India. By the early 1850s he'd been knighted and risen to the rank of general. Sir Harry, accompanied by his devoted Lady Juana, moved to South Africa when he was appointed governor and commander-in-chief. The couple left their mark there too. The South Africans named two cities after them: Harrismith and Ladysmith. Harry died at the age of seventy-three in 1860, and Juana died twelve years later in 1872. They are buried together in Whittlesey.

15

Helen and Edward Thomas

The Last Goodbye, 1917

'I CANNOT STOP CRYING. My body is torn with terrible sobs. I am engulfed in this despair like a drowning man by the sea.'

Helen Thomas's description of her last night with her husband, Edward, is heartbreakingly vivid. It was January 1917. Edward was going to fight in the front line in France. Hundreds of thousands had died there in the first three years of the war. It was the most deadly war zone in Europe and Helen's despair leaps off the page.

'My mind is incapable of thought. Only now and again, as they say drowning people do, I have visions of things that have been – the room where my son was born; a day, years after, when we were together walking before breakfast by a stream with hands full of bluebells; and in the kitchen of our honeymoon cottage, and I happy in his pride of me.'

Helen's husband, Edward Thomas, is remembered as an important twentieth-century British poet. On the horrors of war he wrote:

Any turn may lead to Heaven
Or any corner may hide Hell

But it is Helen's haunting memoir *As It Was* that reminds us just how agonising it is to say goodbye to someone who may never return from the battlefield. She wrote it about her life with Edward as therapy for the grief that consumed her after he died.

Helen was a schoolgirl when she met Edward. In her book she describes herself as 'a plain girl who often cried bitterly in the thought that no man could ever love me'. She fell deeply in love with the grey-eyed young man whose face to her was like 'a portrait of Shelley in its sensitive, melancholy beauty'. The couple married in 1899 while Edward was still a student at Oxford. Helen was already pregnant with their first child.

Their marriage was difficult. Edward suffered from depression and had huge and destructive mood swings. He struggled to make a living from his writing and the family was forced to move many times. He had affairs with other women which he made sure his wife knew about. Helen writes of 'the chequered pattern of our lives', how her husband was often 'bitter and cruel' but she still adored him. In her book she rationalises 'if we love deeply we must also suffer deeply: the price for our ecstatic joy is anguish. And so it was with us to the end.'

The end for Edward Thomas was to be the First World War. He was an early critic of the conflict, publicly condemning British newspapers for being too patriotic. He infuriated his father by writing an article saying the Germans were as brave as the British. When war broke out he was thirty-six, married, and now had three children; so he could have avoided military service but, in the end, he felt he should enlist. His friend the American poet Robert Frost played a part in Edward's change of heart. Frost sent him a copy of his soon to be famous poem 'The Road Not Taken'. Like most of its readers, Edward felt the poem was about indecision and it helped him make up his mind to join the war effort. (Frost later said he was simply poking fun at Thomas who used to agonise about which path to take on their walks together.) Frost had another important impact on Edward, suggesting that he start writing poetry. In the last two years of his life, Edward wrote more than 140 poems.

Edward enlisted in the Artists Rifles in July 1915. At first he was content with a desk job, teaching recruits how to read maps but in 1916 he joined the Royal Garrison Artillery and prepared to fight. In her memoir, Helen describes having terrible premonitions when she learned her husband was going to war: 'An unutterable fear, an icy chill had taken possession of my heart. I was not always conscious of it, but it had the effect of making me feel that life had stopped.'

When Edward announced his regiment was being sent to France early in 1917: 'the icy chill took a closer grip and the sense of statically existing-not-living grew more intense.' Then came joyous news: Edward was coming home for Christmas, to a place Helen calls 'a horrible house on the top of a high hill in Epping Forest, ugly, cold and inconvenient'. But its discomfort was soon forgotten when preparations began for Edward's homecoming.

'Life was not paralysed now but with new-found vigour sped along eager and joyous. Nor did time stand still. I was up half the night arranging the greenery that the children had ransacked the forest for during the day, and the finishing touches to all that was to make this Christmas of all Christmases shine above its peers.'

While the children waited for their father at a nearby crossroads, Helen toasted crumpets for the homecoming tea. Suddenly she heard 'Edward's clear voice calling the old familiar "coo-ee"; then the sound of voices; then of heavy snow-clogged foot-steps, then Edward at the door. He is here. He is home.'

Her extraordinary descriptions of the final days she and her children spent with Edward capture a powerful range of emotions. The bliss of Edward's return, the happiness of Christmas Day, the cheerful visits of friends bringing presents to take to the front are soon replaced by pure anguish. Helen's mood is reflected in her description of snow-covered trees 'which tortured by the merciless wind moaned and swayed as if in exhausted agony. We wearied for some colour, some warmth, some sound, but desolation and despair seemed to have taken up her dwelling-place on the earth, as in our hearts she had entered, do what we would to keep her out.'

The sense of misery grows as Helen relates how she and Edward tried their best to remain positive for the sake of the children. 'We would sit by the fire with the children and read aloud to them, and they would sing songs that they had known since their baby-hood, and Edward sang new ones he had learnt in the army.'

Joining the choruses as she made supper, Helen did all she could to appear cheerful. She watched Edward playing games with the children, 'But he and I were separated by our dread, and we could not look each other in the eyes, nor dared we be left alone together.'

The days passed 'in restless energy'. Edward showed the children how to chop trees into firewood. Helen describes the children helping their father pack his military kit: 'He loved a good piece of leather, and his Sam Browne and high trench-boots shone with a deep clear lustre. We all helped with the buttons and buckles and badges to turn him into the smart officer it was his pride to be.'

Two days before leaving, Edward started sorting through letters and documents on his desk. Knowing he might not come back but not daring to admit it caused a tension that Helen's words capture perfectly:

> I knew Edward's agony and he knew mine, and all we could do was to speak sharply to each other. 'Now do, for goodness sake, remember, Helen, that these are the important manuscripts, and that I'm putting them here, and this key is for the box that holds all important papers like our marriage certificate and the children's birth certificates, and my life insurance policy. You may want them at some time: so don't go leaving the key about.'

When Helen begged her husband to 'leave all this unnecessary tidying business, and put up that shelf you promised me' Edward refused, saying, 'It won't be the first time I've broken a promise to you, will it? Nor the last, perhaps.'

On the last night of Edward's leave he placed a big zinc bath in front of the fire so his daughters could be washed in its warm glow. Helen remembers:

> Edward scrubs them in turn – they laughing, making the fire hiss with their splashing. After the bath Edward reads to them . . . They

> sit in their nightgowns listening gravely, and then, just before they kiss him good night, while I stand by with the candle in my hand, he says: 'Remember while I am away to be kind. Be kind first of all, to Mummy.'

At last, when the children were in bed:

> We are left alone, unable to hide our agony, afraid to show it. Over supper we talk of the probable front he'll arrive at, of his fellow-officers . . . And we speak of the garden and where he wants the potatoes to be, and he reminds me to put in the beans directly the snow disappears. 'If I'm not back in time, you'd better get someone to help you with the digging,' he says.

While Edward read some poetry he'd just written, Helen sat and stared 'stupidly at his luggage by the wall, and his roll of bedding, kit-bag and suitcase. He takes out his prismatic compass and explains it to me, but I cannot see it and when a tear drops on to it he just shuts up and puts it away.'

Edward then read Helen one of Shakespeare's sonnets which leads to one of her most beautiful and poignant descriptions:

> His face is grey and his mouth trembles, but his voice is quiet and steady. And soon I slip to the floor and sit between his knees, and while he reads his hand falls over my shoulder and I hold it with mine. 'Shall I undress you by this lovely fire and carry you upstairs in my khaki greatcoat?' So he undoes my things, and I slip out of them; then he takes the pins out of my hair and we laugh at ourselves for behaving as we so often do, like young lovers . . . Holding the book in one hand, and bending over me to get the light of the fire on the book, he puts his other hand over my breast, and I cover his hand with mine, and he reads from Antony and Cleopatra. He cannot see my face, nor I his, but his low, tender voice trembles as he speaks the words so full for us of poignant meaning. That tremor is my undoing. 'Don't read any more. I can't bear it.' All my strength gives way. I hide my face on his knee, and all my tears so long kept back come convulsively. He raised my head and wipes my eyes and kisses them, and wrapping his greatcoat round me carries me to our bed in the great, ice-cold room . . . Soon he is with me, and we lie

> speechless and trembling in each other's arms . . . Edward did not speak except now and then to say some tender words or name, and hold me tight to him.

For Helen, all Edward's previous misdemeanours were forgiven – his selfish behaviour, cruel words, even his love affairs – when he whispered the following words: 'For no one else but you has ever found my heart, and for you it was a poor thing after all.'

With characteristic honesty, Helen replied: 'No, no, no, your heart's love is all my life. I was nothing before you came, and would be nothing without your love.'

After mutual confessions of love came warm companionship.

> We lay, all night, sometimes talking of our love and all that had been, and of the children, and what had been amiss and what right. We knew the best was that there had never been untruth between us. We knew all of each other, and it was right. So talking and crying and loving in each other's arms we fell asleep as the cold reflected light of the snow crept through the frost-covered windows.

On their last morning together, the children crowded into their bed 'and sat in a row sipping our tea. I was not afraid of crying any more. My tears had been shed, my heart was empty, stricken with something that tears would not express or comfort. The gulf had been bridged. Each bore the other's suffering. We concealed nothing, for all was known between us.'

After breakfast, Edward showed Helen his account books and kissed her when she promised to keep them up to date. He also gave her some poems he had written. Then came their final goodbye.

> We were alone in the room. He took me in his arms, holding me tightly to him, his face white, his eyes full of a fear I had never seen before. My arms were around his neck. 'Beloved, I love you,' was all I could say. 'Helen, Helen, Helen,' he said, 'remember that, whatever happens, all is well between us for ever and ever.' And hand in hand we went downstairs and out to the children, who were playing in the snow.

'Cannon to the right of them, Cannon to the left of them, Cannon in front of them'. Edward Seager survived the charge of the Light Brigade into Tennyson's 'Valley of Death'. Seager commanded a squadron in the third wave of the attack (*right*). Smoke rises from Russian guns firing from three sides

'Bayonet! Forward at a run!' shouts Joshua Chamberlain at the Battle of Gettysburg in 1863. His bold charge leading his troops downhill was the single most decisive moment in the Union's defeat of the Confederate army.

Johnson Beharry (*left*) and his gunner, Troy 'Sammy' Samuels. Their valiant resistance when lethal gunfire hit their Warrior tank in Iraq in 2004 earned Beharry the Victoria Cross.

Laura Secord, her dress tattered after a twenty mile walk, warns British Lieutenant James FitzGibbon that American soldiers are planning a surprise attack in the War of 1812.

Paul Revere's engraving of the 1770 Boston Massacre shows five Americans being killed by British troops. It became a rallying cry for Americans determined to end British rule.

The purple purse used by Harry Percy to carry the Duke of Wellington's dispatch that described the allies' victory over Napoleon's troops at the Battle of Waterloo on 18 June 1815.

'The Meeting of Wellington and Blücher after the Battle of Waterloo', drawn by Daniel Maclise, 1858–9. Harry Percy is the first to the right of Wellington, his face partially concealed.

Krystyna Skarbek happily posing with officers of Britain's Special Operations Executive and French Resistance fighters in Haute-Savoie, France in August 1944.

David Stirling, founder of the Special Air Service in the Second World War, with his men before a raid on Axis forces in the North African desert. Mike Sadler was one of the SAS's ace navigators.

When HMS *Campbeltown* rammed the St Nazaire dry dock gate in 1942, Commando Corran Purdon and his demolition team leapt ashore. Soon after this picture was taken the German onlookers (*left*) were killed when explosives packed in her bow ignited.

Benedict Arnold, the infamous American general, switched sides in the US War of Independence in 1780. He offered to betray the garrison at West Point and kidnap George Washington.

Peggy Shippen, the beguiling daughter of a British judge in Philadelphia, married Arnold. She played a key role in his treacherous plot.

Major John André, the British spy. His furtive mission through American lines to contact Benedict Arnold ended in disaster.

The spy is captured. André is stopped and questioned by three Americans. He protests his innocence until they discover incriminating documents in his boot.

Sonya used the cover of mother and housewife to conceal that she was a Soviet spy. Her son describes this photo, taken in 1936, as 'her perfect clandestine outfit'.

Sonya in 1945 with her children Michael, Peter and Janina in Oxford. At this stage no one knew she had passed Klaus Fuchs top secret information about the Allied nuclear bomb programme to the Soviets.

Ursula Graham Bower took this photo of her Naga 'V' Force scouts on the Indian/Burmese border during the Second World War. They helped her patrol the area, on the look-out for invading Japanese troops.

British troops storm the Bishop's Palace in Badajoz, Spain in 1812.
As marauding soldiers went on the rampage after Wellington's successful siege of the city, Harry Smith rescued the fourteen-year-old Spanish girl he later married.

Afghan tribesmen attack British soldiers and their families fleeing from Kabul in 1842. Lady Sale was wounded and lucky to survive.

Akbar Khan led the rebellion against the British but offered the occupiers safe conduct out of Kabul. The murderous outcome prompted Lady Sale to call him 'the most treacherous of men'.

Crumpled photo of Helen Thomas found in the pocket of her husband, the poet Edward Thomas, after he was killed on 9 April 1917, the first day of the Battle of Arras.

Edward Thomas in 1916 shortly before leaving for France. His wife, Helen, wrote about their heart-breaking final days together.

Alexis Soyer preparing for the grand opening of his first field kitchen in Crimea, 1855. The table on the left will soon be laden with delicacies cooked in his revolutionary Soyer stoves (*right*).

Dr Norman Bethune stands in front of the truck he used in 1936 to transport blood to Republican soldiers wounded during the Spanish Civil War. His invention of mobile blood transfusions saved thousands of lives.

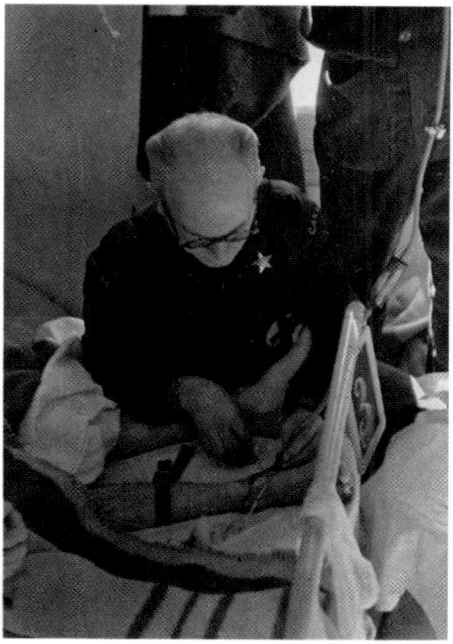

Bethune prepares a wounded soldier for a mobile blood transfusion, Spain, 1937. He watched 'patients' skin turn pink and their eyes sparkle as the bottled blood brought them back to life'.

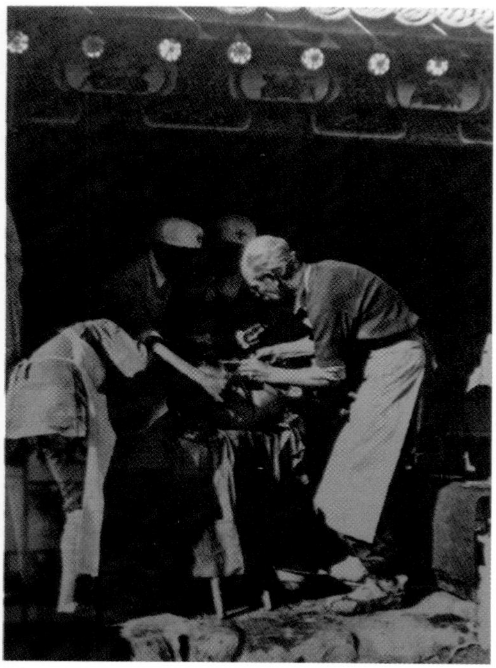

Bethune's make-shift operating theatre in a Buddhist temple on the frontline of the war between China and Japan. He is not wearing sterile gloves because his supply had run out. He died from blood poisoning soon after this picture was taken.

William de Lancey, a senior officer on Wellington's staff fatally wounded at the Battle of Waterloo. His wife of two months, Magdalene de Lancey, nursed him until his death and wrote a memoir so affecting that Charles Dickens said 'I have never seen anything so touching'.

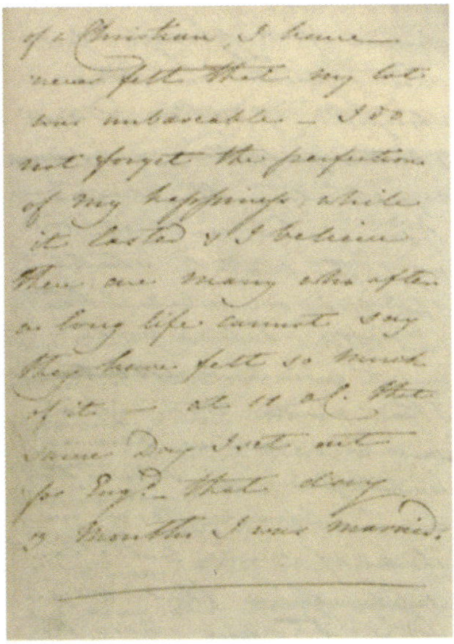

of a Christian, I have
never felt that my lot
was unbearable – I do
not forget the perfection
of my happiness while
it lasted & I believe
there are many who after
a long life cannot say
they have felt so much
of it – at 11 o'c. that
same Day I set out
for Engd. that day
3 Months I was married.

The last page of Magdalene de Lancey's memoir: 'I have never felt that my lot was unbearable. I do not forget the perfection of my happiness while it lasted, and I believe there are many who after a long life cannot say they have felt so much of it.'

Dr Sumner Jackson, the tall man (*centre right*), with his staff at the American hospital in German-occupied Paris where they sheltered many Allied servicemen. His wife, Toquette, is third from the right in the front row.

Augusta Chiwy (*front row, centre*) at her Belgian nursing school, 1943. A year later she treated American soldiers and civilians after Germans surrounded her hometown of Bastogne during the Battle of the Bulge.

American soldiers under German attack at the Battle of the Bulge. Chiwy, just 4 feet 8 inches tall, pulled a soldier to safety after he was shot in the leg.

US General Patton's forces liberate Chiwy's town of Bastogne in January 1945. 75,000 Americans and 80,000 Germans died, were wounded or captured during the Battle of the Bulge.

Captain David Cheap shoots a mutinous midshipman after HMS *Wager* is shipwrecked on the desolate South American coast in 1742.

GOVERNMENT TELLS WELSH MINERS THAT THEY MUST WORK

The Daily Mirror

CERTIFIED CIRCULATION LARGER THAN ANY OTHER PICTURE PAPER IN THE WORLD

WEDNESDAY, JULY 14, 1915

One Halfpenny.

THE E11 RETURNS FROM HER GREAT EXPLOIT IN THE SEA OF MARMORA WITH A HOLE IN HER PERISCOPE.

A front page headline in July 1915 celebrates the astonishing exploits of Martin Nasmith's submarine EII. She dodged Turkish mines and warships in the Dardanelles to reach the enemy capital Constantinople.

British troops wade out to a Royal Navy ship at Dunkirk in 1940. Small boat owners like Charles Herbert Lightoller also saved thousands of lives.

Lightoller's 30 foot boat, *Sundowner*, rescued 129 men from Dunkirk. One soldier was even squeezed into the WC.

Billy Bishop, First World War Canadian fighter ace, stands by one of his earliest aircraft, the Nieuport biplane in August 1917. Few pilots have equalled his claim of 75 German aircraft 'kills'.

Butch O'Hare in his Grumman Wildcat (*behind*) after winning a Congressional Medal of Honour for downing six Japanese warplanes in the Pacific on 20 February 1942.

Franz von Werra's Messerschmitt 109 fighter was a casualty of the Battle of Britain in 1940. He was one of very few Germans to escape from a British prisoner of war camp.

Eric Newby weds Wanda Skof in 1946. She helped him escape German capture in Italy during the Second World War. After the war he returned to Italy to find her.

'Luigi', the farmer who let Eric Newby hide out at his remote farm in the Apennine Mountains in exchange for clearing rocks and entertaining his family with stories about great British murderers.

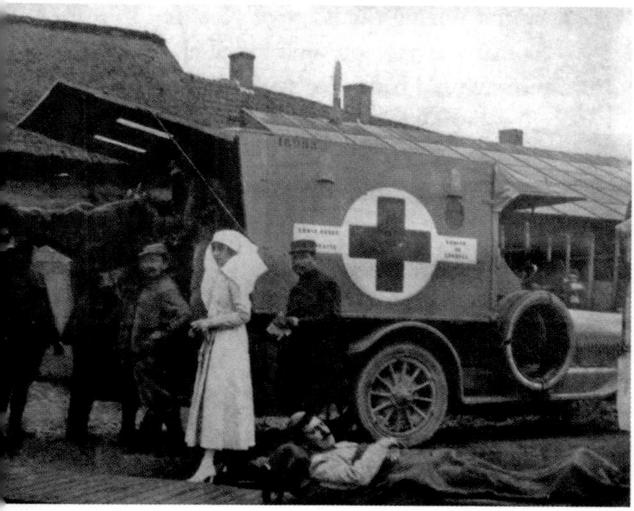

Mary Borden in nurse's uniform outside the First World War hospital she financed in France. It was her job to meet ambulances and decide who needed care most urgently.

Mary Borden (*far left*) chatting with nurses in her mobile field hospital during the Second World War. They travelled with the Free French in Egypt and Libya, often doubling as drivers.

A Japanese soldier about to decapitate a Chinese captive during the Rape of Nanking in 1937–8. As many as 350,000 were killed and thousands more would have died without the valiant efforts of German businessman Jonathan Rabe.

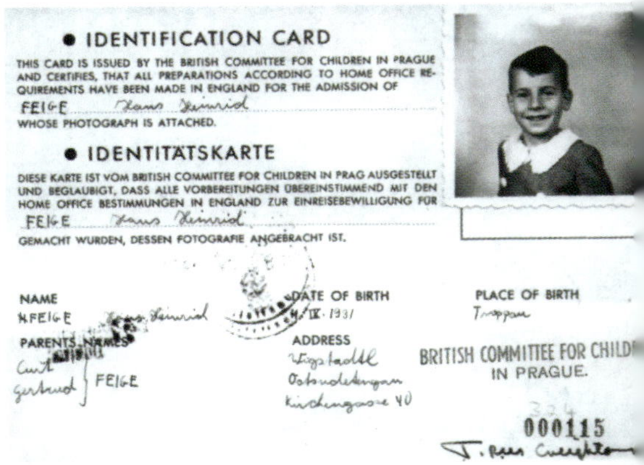

• IDENTIFICATION CARD

THIS CARD IS ISSUED BY THE BRITISH COMMITTEE FOR CHILDREN IN PRAGUE AND CERTIFIES, THAT ALL PREPARATIONS ACCORDING TO HOME OFFICE REQUIREMENTS HAVE BEEN MADE IN ENGLAND FOR THE ADMISSION OF

FEIGE Hans Heinrich

WHOSE PHOTOGRAPH IS ATTACHED.

• IDENTITÄTSKARTE

DIESE KARTE IST VOM BRITISH COMMITTEE FOR CHILDREN IN PRAG AUSGESTELLT UND BEGLAUBIGT, DASS ALLE VORBEREITUNGEN ÜBEREINSTIMMEND MIT DEN HOME OFFICE BESTIMMUNGEN IN ENGLAND ZUR EINREISEBEWILLIGUNG FÜR

FEIGE Hans Heinrich

GEMACHT WURDEN, DESSEN FOTOGRAFIE ANGEBRACHT IST.

NAME FEIGE Hans Heinrich

DATE OF BIRTH IX. 1931

PLACE OF BIRTH Troppau

PARENTS' NAMES Curt, Gertrud FEIGE

ADDRESS

BRITISH COMMITTEE FOR CHILD[REN] IN PRAGUE.

000115

The identification card of seven-year-old Hans Heinrich Feige. He was saved from the Nazis by Nicholas Winton's Kindertransport in 1939. His Jewish parents remained in Czechoslovakia and died in Auschwitz.

Ahmad Terkawi's city, Homs, was devastated in the fighting in Syria that began in 2011. With his pharmacy in ruins he fled the country. He and his wife had to dive into the Aegean an rescue their two small children after people smugglers threw them overboard.

Edward asked the disconsolate Helen to remain in the house. The children would walk him to the station. 'A thick mist hung everywhere, and there was no sound except, far away in the valley, a train shunting. I stood at the gate watching him go; he turned back to wave until the mist and the hill hid him.'

The last words Helen Thomas heard from her beloved husband Edward as he walked off to join his regiment in January 1917 was a faint 'Coo-ee!' During their seventeen-year marriage, the couple joyfully shouted this greeting as they approached or left each other. But on this day the sound was heart-breaking.

> I heard his old call coming up to me. 'Coo-ee!' he called. 'Coo-ee!' I answered . . . Again through the muffled air came his 'Coo-ee!' And again went my answer like an echo. 'Coo-ee!' came fainter next time . . . I put my hands up to my mouth to make a trumpet, but no sound came. Panic seized me, and I ran through the mist and the snow to the top of the hill and stood there a moment dumbly, with straining eyes and ears. There was nothing but the mist and the snow and the silence of death.

Just three months after he left home Edward Thomas died on Easter Monday, 9 April 1917. He was killed on the first day of the Battle of Arras, a five-week struggle on the Western Front during which allied armies failed in yet another attempt to end the stalemate with German troops. Some 160,000 allied soldiers were killed or wounded in the fighting. There were 125,000 German casualties. Edward's commanding officer informed Helen that shockwaves from an exploding shell stopped both her husband's heart and his watch at the same time. She wrote to a friend: 'He told me there was no wound and his beloved body was not injured.'

The recent discovery of a letter from the same commanding officer revealed that Edward was actually 'shot clean through the chest'. The story about his heart stopping was designed to spare his distraught widow any further suffering. Edward Thomas is commemorated in Poets' Corner in Westminster Abbey. His wife Helen's memory lives on in the powerful, passionate memoir she wrote as a grieving war widow.

INNOVATION

INNOVATION WINS WARS. Britain's tank helped the allies in the First World War to take the battle out of the trenches and on to victory over Germany in 1918. America's atomic bomb obliterated Nagasaki and Hiroshima and forced the surrender of Japan. The last century has seen military technology leap ahead at a frantic pace as nations compete to match or outdo each other's weapons development.

But war has stimulated a less destructive kind of creativity too. When we started researching this book we were struck by how many things we take for granted in our daily lives come from military inventiveness. That cup of instant coffee we gulp down in the morning is a direct descendant of the coffee concentrate first used by soldiers in the American Civil War. We have Napoleon Bonaparte to thank for canned food. In 1809 as he attempted to conquer Europe, the French government organised a competition to discover the most efficient way to get food to front-line troops. The winner, Nicolas Appert, proved he could preserve food by putting it inside a sealed glass jar which he then heated up. Germ-free, it could be stored for months and was easily transported. The zippers on our clothes were first used to fasten uniforms in the First World War. Other innovations from that war include face tissues whose original purpose was to line gas mask filters and sanitary napkins, a direct descendant of an exceptionally absorbent material first used as surgical dressings. The computers we use today owe their origin to Colossus, the complex machine invented to break top-secret German codes during the Second World War. Jet engines and nuclear power are just two of a long list of technological advances that began in the Second World War.

Then of course there are the medical innovations. The first modern plastic surgery took place in 1917 when skin was grafted over the face of British sailor Walter Yeo whose eyelids had been burned off during the Battle of Jutland. At the same time in France, the Nobel Prize-winning radiologist Marie Curie was setting up newly invented mobile X-ray machines. They travelled to field stations in cars and trucks known as 'little Curies'. Penicillin was discovered by Alexander Fleming in 1928 but it was only in the Second World War that it was put into mass use.

Two wartime innovators stand out for us: one was a pioneer in the field of medicine and the other a culinary genius. Dr Norman Bethune developed portable blood transfusions during the Spanish Civil War, then went on to create mobile operating theatres for Mao Zedong's Red Army in China. But we begin this chapter with the enthralling story of Alexis Soyer, a celebrity cook who changed the way food was prepared and cooked for British soldiers nearly 100 years earlier in the Crimean War.

16

Alexis Soyer

Celebrity Chef, Crimean War, 1855–6

On 2 February 1855, Alexis Soyer, Britain's most famous chef, left a London theatre to join friends at a nearby restaurant. A waiter showed him to the wrong room and while waiting in vain for his fellow diners, he picked up a copy of *The Times* newspaper and read the latest distressing report from the front line of the Crimean War. *The Times* had sent its reporter William Howard Russell to the war-ravaged peninsula and, taking advantage of the newly invented telegraph, he sent back the first eyewitness reports from a battleground. Many of us who have covered news stories regard Russell as the first serious war correspondent. His chronicles described the dreadful conditions facing British soldiers on the battlefield and in hospitals. They shocked the British public and forced the government to change the way it supplied and treated its fighting forces.

Russell wrote about incompetent British military commanders, of soldiers dying in filthy hospitals, and of poor food supply. He described men 'enfeebled by sickness' and 'hungry and wet and half-famished'.

His writing exposed the fact that the army was shamefully neglecting its soldiers.

Deeply moved by what he had read, the tender-hearted chef asked for a pen and paper and wrote a letter to the editor of *The Times*. Soyer offered to travel to the war zone at his own expense to ensure the troops received properly prepared and nutritious food 'acting according to my knowledge and experience in such matters'. Russell's dispatches about woefully poor medical care in Crimea had already inspired *The Times*'s readers to start a fund that sent Florence Nightingale and a team of nurses to the area.

The Crimean War that aroused such strong feelings in Soyer and others began in 1853 when Russia attacked Turkish territories in the Balkans. Britain and France went to Turkey's aid and their forces landed in Crimea, well behind the Russian front line. They met with some immediate success but within months the war was bogged down and casualties on both sides mounted rapidly. Along with incompetent military commanders, ghastly mistakes like the Charge of the Light Brigade and appalling medical care, feeding the troops became a major issue. The British army's food supply authorities, known as the Commissariat, were notoriously inept and corrupt. British and French army caterers also bid against each other for local produce, pushing prices sky high.

The majority of soldiers who died in Crimea perished not from war wounds but from sickness often caused by grossly substandard food provided by unscrupulous suppliers. Working closely with Nightingale, the celebrity chef from London had a profound influence not just on how food was prepared and served in the Crimean War but on future conflicts too.

Soyer had moved to London from his native France in 1831. Although just twenty-one, he'd already been one of the French Prime Minister's personal chefs. In England he rose to fame as head chef of the new Reform Club founded by leading Liberals. An energetic innovator, he introduced gas cookers, water-cooled refrigerators and ovens with adjustable temperatures to the club's state-of-the-art kitchens. Soyer, a flamboyant but charming self-promoter, was easily recognisable as he strode around London buying food, doing deals and planning new ventures. Enveloped in a weirdly shaped cloak, he wore a trademark sloping hat and carried a slanted cane. His eccentric attire gave rise to much comment. It was noted that his clothes were cut on the bias or diagonally – something the chef described as *à la zoug-zoug*.

Soyer loved being in the limelight and was in his element preparing outrageously elaborate dishes for his aristocratic patrons. His *Chapons à la Nelson* featured chickens cooked in pastry shaped like the prow of a ship, floating on a sea of mashed potato.

He may have catered for the rich but he also had a strong social conscience. Soyer set up soup kitchens in Ireland during the great famine, serving a nutritious beef and vegetable broth called 'Soup for the Poor'. *Punch*, the satirical magazine, was not impressed and labelled his concoction 'Poor Soup'. He wrote cookbooks full of inexpensive but healthy recipes and offered his services free of charge to poorhouses and hospitals, making their kitchens and food production more efficient and economical.

When Soyer wrote his letter to *The Times* offering to travel to the Crimea, the British government could not believe its good fortune. A celebrity chef, a household name, was volunteering to sort out two of its most pressing failures: filthy kitchens in military hospitals and ill-prepared food in the field.

Alexis Soyer was not only famous for his cooking, he was also an inventor. Two of his most popular designs were a vegetable steamer and a clock that rang when food was ready. Lord Panmure, Britain's Minister of War, quickly summoned Soyer to a meeting. It was agreed that he would come up with a new invention: a field stove to replace the outdated tin kettles used by soldiers to cook meals. Soon to be known as Soyer's Stove, it would revolutionise the way food was prepared for British soldiers. The stove resembled a rubbish bin perched on a burner. On top of this contraption there was room for a large cauldron, which could hold enough to feed fifty people – eight times as many as the tin kettles currently in use. The new stove also required far less wood than the open fires needed to heat kettles. Soyer calculated his invention could save an army of 40,000 men 90 tons of fuel a day. As well as taking his stove to the battlefields, Soyer was also asked to improve soldiers' diets. Lord Panmure flippantly urged him to 'go to Crimea and cheer up those brave fellows in the camp. See what you can do. Your joyful countenance will do them good, Soyer: try to teach them to make the best of their rations.'

So a month after writing his letter to *The Times*, Soyer was on his way to Constantinople. When he stopped at Marseille, he came face to face for the first time with the grim reality of the war. He described seeing '700 or 800' men who had just landed from Constantinople and Crimea. 'Their appearance, I regret to say, was more than indescribable . . . those who were wounded looked joyful compared with those who were victims of epidemic – typhus fever, diarrhoea, dysentery, cholera or frostbite.'

Another horrifying sight greeted Soyer in March 1855 when he arrived in Scutari, a suburb of Constantinople. Florence Nightingale took him on a tour of the Barracks Hospital, one of six centres that treated wounded soldiers. Nightingale was far more than 'the lady with a lamp'. She transformed military hospitals by cleaning up treatment areas and introducing her own trained medical staff. But during their tour Soyer noticed that the hospital kitchen was filthy. Rats ran rampant. Cooking was done by untrained soldiers who served in rotation and could hardly wait to get back to their normal duties. Soyer wrote: 'the hellishly hot and smoky conditions in the kitchens – exacerbated by the indiscriminate burning of whole trees, leaves and all, to keep the copper furnaces fired up and the water boiling fiercely meant the job was despised.' He watched in disbelief as orderlies tied joints of meat to wooden paddles and threw them into boiling water. They identified their meat by attaching an object to it – everything from 'a string of buttons, a red rag, a pair of surgical scissors or even, in one case, some ancient underwear'. When Soyer told one cook, 'it was a very dirty thing to put such things in the soup', the sweating cook answered: 'How can it be dirty, sir? Sure they have been boiling this last month.'

Soyer leapt into action. The kitchen was scoured clean and metal skewers replaced the wooden paddles. Cooking fuel was used more efficiently to reduce the heat and smoke. Soyer's clearly written recipes appeared on kitchen walls and cooks learned how to prepare simple dishes. He provided his own recipes for nutritious food for patients like mutton and barley soup and calves'-foot jelly which you could wash down with 'Soyer's cheap Crimean lemonaid'. He taught the cooks and orderlies how to recycle food. Cooking water that had previously been thrown out was now used to make soup. Fat from the water became a substitute for butter.

He urged the army to end the system of using soldiers to do short-term stints in the kitchens. It took Soyer's advice and hired civilian cooks. He went on to clean up other British military hospitals but still managed to find time to come up with a new invention, the Scutari teapot. Until his arrival, cooks made tea by dumping tea leaves wrapped in a cloth into kettles that had been used for making soup. The watery result tasted more of broth than tea. By putting the leaves into a strainer on top of a kettle, Soyer found 'to my astonishment it made about one-fourth more tea, perfectly clear and without the least sediment'.

Soyer wrote regular letters to London publications trumpeting his achievements. Queen Victoria got her own first-hand report in a letter from Lady Stratford, wife of the British ambassador in Constantinople: 'M. Soyer has done much good in the kitchens. He is a most ridiculous man but quite perfect in his way.'

Two months after his arrival in Scutari, the energetic Soyer was off again. The Light Brigade's Edward Seager wrote in a letter home: 'Soyer goes to the Crimea this week and I hear Miss Nightingale accompanies him for a short visit. He is going to teach the men how to cook their rations so as to make a palatable meal. His cookery here is perfection. He is much liked for his affable and gentlemanly manners.'

Soyer had hoped to take 400 of his stoves with him to Crimea but they had not yet arrived from England, so the chef set off on 2 May 1855 with just ten. Wearing a flamboyant red and white turban and a hooded cloak on board a large troopship, the *Robert Lowe*, he enjoyed the company of Nightingale whom he described as 'amiable and gentle'. They arrived in Balaclava in the south-west tip of Crimea, now securely established as the supply base for the allied siege of Sebastopol. As the duo travelled from hospital to hospital they despaired of the uncomfortable and dirty conditions they found. Many of the kitchens were made of mud and had no roofs, so Soyer designed suitable wooden structures. He also created two new nutritious foods for soldiers in the field. The first was a vegetable cake containing dried carrots, leeks, turnips, parsnips, cabbage, celery and onions made tasty with seasonings. Then there was a bread biscuit made from flour and peasemeal. 'It will keep for months,' he said, 'and then soak well in tea, coffee or soup.'

Travelling around hospitals and kitchens in French and English camps near the besieged city of Sebastopol, Soyer marvelled at the 'myriads of white tents, the sound of trumpets, the beating of drums, the roar of cannon from Sebastopol'. Once, he and Nightingale decided to 'take a peep' at the city. They stopped at a battery in a trench and Soyer induced Nightingale to sit on a cannon. 'I then boldly exclaimed "Gentlemen, behold this amiable lady sitting fearlessly upon that terrible instrument."'

In spite of his very busy schedule, Soyer, as usual, made time for socialising. He became a regular at the British Hotel run by the redoubtable Mary Seacole, a Creole nurse from Jamaica. Her café, bar, shop, restaurant and hostel was also a clinic. Although Nightingale would later claim the British Hotel was little better than 'a bad house', another word for a brothel, Soyer recognised Seacole as a kindred spirit, describing her as 'good and benevolent'. She dubbed him 'the great high priest in the mysteries of cookery', and wrote: 'There was always fun when the good-natured Frenchman was there.'

It was certainly fun when Soyer's stoves finally made it to Balaclava in August 1855. Hundreds of British and French soldiers and doctors were invited to attend a grand opening of the first field kitchen with Soyer putting his invention through its paces. The stoves were carefully arranged around white-clothed tables laden with wines and champagne. There was even a band. The menu included 'plain boiled salt beef; ditto, with dumplings; plain boiled salt pork; ditto, with peas-pudding; stewed salt pork and beef, with rice; French pot-au-feu; stewed fresh beef, with potatoes; mutton ditto, with haricot beans; ox-cheek and ox-feet soups; Scotch mutton-broth; common curry, made with fresh and salt beef'. One soldier wrote home that Soyer 'certainly made very nice ragouts and soups, but I fear it will be a very long time before we can do it for ourselves'.

Soyer did not get the chance to see his stoves being used on the front lines because shortly after his grand opening, the allies captured

Sebastopol. Around 23,000 people on both sides were killed and wounded in the final assault on 8 September 1855. Soyer volunteered to help in the kitchens of the general hospital where he saw so many amputations that 'several buckets' were filled with limbs. He and a friend went into the now ruined city and Soyer headed straight for the Russian kitchens. He dismissed their cabbage soup as 'extremely bad and entirely deprived of nutritious qualities'. Even in the midst of such destruction, he could not resist playing a joke on his friend. They both took a bite of black bread then Soyer pretended to gag, shouting it had been poisoned. He gleefully reported that his friend put 'his fingers in his throat in order to throw off the dreadful meal' while Soyer roared with laughter.

What happened next was far from a laughing matter. A week after the fall of Sebastopol Soyer was struck down by Crimean fever, a bacterial infection caused by bad meat or unpasteurised milk. It was often fatal. Nightingale had caught the disease two months earlier and barely survived. Soyer spent weeks in bed with severe fever, heart palpitations, headaches and insomnia. He was at last able to get up looking 'so altered that scarcely anybody could recognise me', and followed his doctor's advice to return to Constantinople to convalesce. There he was laid low by dysentery but still decided to make one more trip to Balaclava. Fifty of his stoves had arrived in Constantinople and, even though the war was over, he wanted to make sure they reached soldiers who needed them. When he ignored medical advice not to travel, his exasperated doctor warned, 'Don't forget to take your gravestone with you.'

Once back in Balaclava, Soyer arranged for stoves to be delivered to each of the forty regiments still in the camp. Rising at six each morning and often working twelve-hour days, he visited every regiment explaining how to use the stoves. He also handed out simple recipes in the hope that the soldiers would master them before being sent home. Unsurprisingly, Soyer somehow found time to entertain. Not content with accepting numerous invitations, he held dinner parties of his own complete with dishes like his special Tally-ho Pie, from which a live fox leapt out when the pastry was cut. A dessert called *La bombe glacée à la Sebastopol* was a big hit as was Crimean Cup *à la Marmara*, a heady mixture of champagne, rum, lemon and sugar. With the war over he said the camp resembled a 'monster banqueting hall'.

Soyer left the Crimea for Constantinople on 10 July 1856. A shameless self-promoter, he gathered testimonials from influential people. He asked Florence Nightingale for a 'candid opinion of the humble services I have been able to render to the Hospitals'. She had been critical of him in the past, calling him a 'humbug', but this time she sent a glowing report saying he had 'restored order where all was unavoidable confusion' and 'took soldiers' and patients' diets and converted them into wholesome and agreeable food.' As for his stoves, she wrote they 'answer every purpose of economy and efficiency'.

Soyer arrived back in London in May 1857 after several months travelling in Europe. He had not fully recovered from his Crimean fever and doctors told him to rest. Ignoring their advice, he 'ran on in a mad career of gaiety'. His inability to say no to invitations caused further problems. One day when riding to join friends for lunch, his horse bolted. Soyer's foot got caught in the stirrup and he was dragged down the road before the horse was stopped. Nothing was broken and in spite of feeling understandably shocked, he went on to the lunch in a hansom cab.

The accident and fast-failing health did not stop him inventing a new sauce with Turkish herbs. He wrote another cookbook with the ungainly title of *A Culinary Campaign, being Historical Reminiscences of the late War with the Plain Art of cookery for Military and Civil Institutions, the Army, Navy, Public, etc.* He gave a lecture at the United Service Institute on military and naval cookery. Ever the showman, he ended his talk by producing delicious soup and omelettes on one of his Soyer stoves. Nightingale asked him to design a model military kitchen and to teach army and hospital cooks new recipes. He started work on a new stove which could cook 'a dinner either for one man or a battalion'.

By the summer of 1858, just forty-eight years old, Soyer was paying the penalty for his frantically full life. He was spitting blood and losing weight. Drinking copiously, his behaviour became bizarre. He shouted at servants, then according to one witness, 'he would dive into stew-pans and kettles.' This extraordinary behaviour was consistent with Crimean fever. The end was not long in coming. Driven as ever, he was designing a mobile cooking carriage for the army when he fell into a coma and died.

The *Morning Chronicle* wrote: 'he saved as many lives through his kitchens as Florence Nightingale did through her wards.' The great Nightingale herself was moved to comment: 'Soyer's death is a great disaster . . . He has no successor.'

In spite of his short life, Soyer's legacy is immense. The Crimean War led to the Soyer Stove becoming part of the English language along with the cardigan and the raglan sleeve. Soyer's Stoves were used for over a century by armed forces in Britain, Canada and Australia. They provided hot meals and tea in British cities bombed during the Second World War. They even went to the Falklands War in 1982. Soyer's insistence on properly trained army chefs led to the creation of the Army Catering Corps, while his soup kitchens became a model for charities such as the Salvation Army and Oxfam. His cookbooks, with their emphasis on nutritious, tasty and economical food, inspired culinary stars like Mrs Beeton, Elizabeth David and Jane Grigson. His signature Lamb Cutlets Reform are still being served at the Reform Club just as they have been since 1846.

17

Dr Norman Bethune

Pugnacious Medical Genius, 1936–9

IN NOVEMBER 1936 two Canadian men drove through the Rhône Valley in France heading for the Spanish Civil War, which was part of the great conflict between the forces of liberal democracy and Fascism. On the side of their battered Ford station wagon, a handwritten sign made it clear the Canadians – like the majority of foreign volunteers – were on the side of the socialist Republican government of Spain against a rebellion by General Franco's Fascists. French sympathisers raised clenched fists in support as the car drove by.

One of the men, Dr Norman Bethune, barely noticed. He spent much of the journey reading a book about blood transfusions. A physician from Montreal, he was a medical pioneer on the verge of a major breakthrough. Transfusion was not new but it was a very inflexible process. The donor had to be present, beside the recipient. Blood could not be preserved.

Bethune came up with a solution. He perfected a process that allowed donated blood to be stored then moved in bottles to the front line. His revolutionary blood transfusion service saved the lives of thousands of soldiers who previously would have bled to death where they fell. Norman Bethune had not come to Spain to fight but to heal.

Four months before Bethune arrived in Madrid, General Francisco Franco had declared war on the recently elected Republican government of Spain. He was supported by the establishment – rich landowners, big business, leaders of the Catholic Church and the Falange, Spain's Fascist party. European Fascists also rushed to Franco's aid: German leader Adolf Hitler lent him 100 war planes and 16,000 troops; Italy's Benito Mussolini provided 50,000 soldiers. The Spanish government was backed by workers, liberal writers and artists and left-wing political parties along with 40,000 volunteers like Bethune from more than sixty countries.

Bethune had always thrived on taking chances. In 1896, when he was just six years old, he disappeared from his Toronto home one morning, returning hours later to announce that he had walked fifteen kilometres across the city. His father, a Presbyterian minister, was not amused. Bethune loved being outdoors and tackling new challenges. He dropped out of university for a year to work at a logging camp in northern Ontario. After chopping down trees all day, he spent the evening teaching lumberjacks how to read and write. He wanted to be a doctor but his medical studies were interrupted by the First World War. Serving as a stretcher-bearer at the Battle of Ypres gave him first-hand knowledge of how quickly the wounded died from lack of blood.

After the war Bethune completed his medical studies and headed to London, a city he loved for its art galleries, fancy restaurants and excellent tailors. To pay for his extravagant lifestyle the trainee doctor sold antiques. He worked at the Hospital for Sick Children on Great Ormond Street and became fascinated by surgery. In 1923 he married a Scottish heiress, Frances Campbell Penney, whom he later divorced, married again and divorced a second time – all within ten years.

At this stage of his life, friends described Bethune as 'slapdash', 'refreshing' – a handsome man who loved an argument and enjoyed life to the full. Women loved him. One described the dashing doctor as 'the most aggressively male creature I ever encountered'.

Back in North America, his zest for life was severely tested in 1926 when he was struck down by tuberculosis. He went to a sanatorium in New York State for treatment. He had fun there at first, sneaking out of his room at night and drinking bootlegged whiskey in a nearby town. But as months passed and his disease got worse, Bethune feared he was dying. He thought of suicide and painted a mural that included his gravestone. Fortunately the sanatorium had a good library and he came across an article in a medical journal that gave him hope. It described a procedure called pneumothorax where a needle is used to collapse a diseased lung in order to let it heal. Bethune rushed from the library to his doctor, gave him detailed instructions about how to perform pneumothorax and demanded to go straight to the operating theatre. The operation was so successful that when the surgeon checked on Bethune hours after collapsing his lung, he found him standing on a chair finishing his mural.

His brush with death changed Bethune dramatically. He later wrote that he'd had time to think during his year in the sanatorium and that this had deepened his intellectual and spiritual life. Being treated at one of the best and most expensive sanatoriums in North America had also awakened Bethune's social conscience. He was appalled that 'there are two kinds of tuberculosis – the rich man's and the poor man's. The rich man lives and the poor man dies.'

He told friends he was determined to do something great for the human race before he died. Finding a cure for tuberculosis became Bethune's mission. He took up thoracic surgery in Montreal where he proved to be a talented surgeon but an extremely difficult colleague. He condemned other surgeons' old-fashioned ways, shouted at nurses and complained about equipment. One doctor remembered seeing a trail of discarded surgical knives Bethune scattered along a hospital hallway because they were not sharp enough. His rejection of outdated practices and equipment made Bethune unpopular. He was undoubtedly tiresome but his perfectionism brought huge benefits to his profession. He invented new surgical instruments that transformed chest surgery. His Bethune Rib Shears – described in a catalogue as having 'long handles, powerful biting jaws, suitable for all ribs' – are still in use today.

Outside the hospital the combative doctor hung out with artists, painted, wrote poetry, plays and short stories. He was a vain man, proud of his good looks. Though a self-proclaimed socialist, he enjoyed life's luxuries, splashing his face with the most expensive colognes and dressing elegantly. In 1935 he travelled to Russia where he was hugely impressed by its free hospitals and on his return joined the Communist Party of Canada, a banned organisation. He also became a passionate campaigner for free health care for all Canadians. He urged the Canadian government to set up a socialised health system but his plans were rejected. So when the Spanish Civil War started, Bethune could hardly wait to leave Montreal and rush to the aid of his Spanish comrades. He looked on Fascism as a disease, an insanity that threatened civilised society: 'It is in Spain the real issues of our time are going to be fought out. It is there that democracy will either die or survive.'

Once in Madrid, he wrote to a friend: 'The world war has started – democracy against fascism. I wouldn't be anywhere else. This is the hub of the universe.' In November 1936 Madrid was a city under siege 'like a hunted animal' according to Bethune. Franco's troops surrounded most of the Spanish capital. He wrote that Fascist bombs fell like 'great black pears leaving bodies where once there were women and children'.

Bethune quickly set up the Servicio Canadiense de Transfusion de Sangre. Scientists had been experimenting with blood transfusions for centuries. The first attempts took place in France in 1667 when King Louis XIV's doctor transferred animal blood into humans with disastrous results. Transfusions then fell out of fashion until the nineteenth century when the invention of rubber tubing allowed blood to be carried from donor to recipient. In most cases, however, the donor had to lie on a bed attached by tubes to the person who needed the blood.

The Canadian doctor's ingenuity now raised the science of blood transfusion to a new level. Once blood was taken, he and his workers put it into a bottle and stored it in a refrigerator. Crucially, they added sodium citrate to stop it clotting. In order to build up his blood bank, Bethune took the unusual step of appealing for donors on radio and in newspapers. The result was extraordinary. Republican patriots queued up in their hundreds, so many that on some days it was not possible to take blood from everyone in line. When any of the more than fifty hospitals in Madrid requested blood, Bethune's unit made sure they got it.

Then in February 1937 Bethune took his transfusion unit to the battlefront for the first time, transporting the blood in trucks. He and his mobile transfusion team covered a huge area containing more than 100 hospitals and first-aid stations. As soon as a call for blood came, the refrigerator would be loaded and Bethune and his driver would be off. He later spoke of seeing wounded soldiers lying white-faced and motionless in their hospital beds. He described how he watched their skin turn pink and their eyes sparkle as the bottled blood brought them back to life.

It was dangerous work. At one point he and his team were forced to jump out of their truck when it came under fire from Franco's soldiers. On another occasion while Bethune was taking his mobile blood unit to southern Spain he witnessed a scene that would haunt him for the rest of his life:

> When the little sea port of Almeria was completely filled with refugees, we were heavily bombed by German and Italian Fascist airplanes . . . They deliberately dropped ten great bombs in the very centre of town on the exhausted refugees . . . After the planes had passed I picked up in my arms three dead children from the pavement . . . In the darkness the moans of wounded children, shrieks of agonized mothers, the curses of the men rose in a massed cry higher and higher to a pitch of intolerable intensity.

Bethune came away from this horror more determined than ever to help the victims of war and to fight Fascism.

Twenty-four-hour days dodging enemy fire took its toll. Bethune often drank too much and was not easy to work with but his charisma still worked wonders. According to one of his staff: 'He was literally the sort of person who could say "rise up and follow me" and you would follow him. He had that sort of authority about him.'

After seven months Bethune fell out with Spanish authorities who wanted to take over his transfusion service. He decided to return to Canada, leaving behind a fully functioning mobile blood service. It's estimated that the number of people who died from their wounds dropped by 30 per cent thanks to Bethune's inspired pioneering work. He now moved on to employ his life-saving inventiveness in a completely different way on the other side of the world.

Bethune was on a cross-Canada tour, raising money for the Spanish Republican cause, when on 7 July 1937 Japan launched a full-scale attack on China. China was already in turmoil with the Nationalists under General Chiang Kai-shek attempting to destroy Communist forces led by Mao Zedong. The Communists had been pushed back but were still fighting from a base at Yan'an in the north-west of the country. It was only when it became clear that China was in danger of being overrun by Japan that Mao signed a truce with his former Nationalist enemies. Bethune believed the Chinese were involved in the same struggle as Spain – a fight against Fascism – only on a much greater scale. He resolved to help Mao. In January 1938 he left Vancouver on a ship headed for Hong Kong, accompanied by two other Canadians, Dr Charles Parsons, an experienced doctor but an alcoholic, and a nurse, Jean Ewan. Unlike her travelling companions, Ewan could speak Chinese. A brave, no-nonsense woman, she came to loathe the short-tempered Dr Bethune. When they arrived in Hong Kong it emerged that Dr Parsons had spent most of their expenses money buying drinks on board the ship. There was nothing left to pay for hotel rooms. Bethune hit the roof. Nurse Ewan wrote: 'I have not seen such a temper before, except in my father. He stomped and kicked everything in sight except Parsons.'

Parsons was left behind as Bethune and Ewan made the dangerous journey to join Mao's Red Army. It took them six weeks by foot and train to travel 1,300 kilometres to Mao's headquarters in Yan'an. Along the way, Bethune treated wounded soldiers and civilians in old huts, barns – anywhere that was sheltered from Japanese bombs. Nurse Ewan helped him remove bullets from arms and legs, set broken bones and remove infected limbs. Even in the midst of death and destruction, Ewan had it drummed into her how Bethune expected her to behave: 'I was never to call him by his first name. I was not to take it upon myself to diagnose or treat patients. I was a servant, no more, no less.'

Ewan found Bethune unsympathetic and rude but she was impressed by how calmly he accepted the dangers they encountered on their journey. They were bombed, shot at and nearly killed several times. On one occasion when Japanese troops were less than ten miles behind them Bethune's only concern was treating the wounded. She admitted that he was a gifted physician who had a great love for humanity and assumed that somewhere under his prickly exterior there was a soft heart. Later she said she wondered even about that.

When they finally arrived in Yan'an, the whole town was waiting, waving flags and banging drums for their honoured guests, since Dr Bethune's reputation had reached even this out-of-the-way place. No sooner had he and Ewan collapsed into their beds than a knock at the door summoned them to meet Mao, the Communist leader. Their escort explained that Mao worked at night, from midnight until eight or nine in the morning because that was when it was quiet. He didn't usually see people unless they were important. The two Canadians were taken to a cave where Mao lived. In the candlelight Ewan recalled seeing books and papers piled on his table. She was struck by how long, soft and sensitive the Chinese leader's hands were. He greeted his guests with a smile and embraced Norman Bethune. Drinking tea and eating peanuts, they talked for hours about the wars in China and in Spain. They also discussed the need for good medical care for Mao's soldiers and how Bethune's profile resembled Lenin's.

It was decided that Bethune's medical skills were most needed at the front, several hundred kilometres to the north. Nurse Ewan was sent off to buy medicine but was unable to make her way back. She never worked with Bethune again. He may not have missed her but he could have used help. He wrote to friends: 'In this great area of thirteen million people, and 150,000 armed troops, I am the only qualified doctor. The Chinese doctors have all beat it. My life is pretty rough and sometimes tough as well. It reminds me of my early days up in the Northern bush.'

His experience in China was very different from Spain. There he had sterile laboratories to work in, a reasonable flow of medical supplies and well-trained staff to support him. In China, conditions were unhygienic and medicine was scarce so he had to improvise to save lives. His most important invention was a mobile medical unit, which travelled in wooden cases that fitted on the back of a donkey. The cases contained a folding operating table, twenty-five wooden legs and arms to replace severed limbs and supplies for 100 operations. The staff accompanying these units could go into action within thirty minutes of arrival.

Bethune spent his days frantically treating patients whom he described as having 'old neglected wounds of the thigh and leg – most of them incurable except by amputation . . . They are without exception, all anaemic, underfed and dehydrated. They are dying of sepsis.' On one occasion he performed seventy-one operations in forty hours with the help of two assistants he had trained. Training became a major part of his work. When he wasn't looking after the sick and dying, he gave classes in basic medical care. He even found time to write a textbook. The medical instruments Bethune had brought with him from Canada outnumbered the entire supply in Mao's army. If an instrument broke or went missing he was quick to invent a replacement, just as he had years before in Montreal. He built an operating room in a temple, designed splints for legs and arms and created a steriliser.

Working long hours constantly on the move, sleeping in peasant huts full of lice, and getting frostbite had a disastrous effect on Bethune's health. He lost thirty pounds in weight and looked much older than his forty-nine years but told friends he hadn't been so happy for a long time. In less than two years Bethune became a hero in China. He was nicknamed Bai Quiuen, which means White One Seeking Grace. Mao's party spread word of the Canadian doctor's work throughout the land and soldiers chanted this slogan before attacking the enemy: 'We fight at the front. If we are wounded, we have Bai Quiuen to treat us. Attack!'

On 28 October 1939 Bethune cut his finger during an operation. His immune system was weak so the wound remained open. In early November he performed brain surgery on a soldier. He had run out of gloves so his cut came into contact with the man's highly infected brain tissue. Bethune developed a fever, his finger became swollen and an abscess appeared in his armpit. On the morning of 12 November

1939 Norman Bethune died in a tiny hut in a Chinese mountain village.

He left a note saying he knew he was dying and his only regret was that he was not able to do more. The letter ended: 'The last two years have been the most significant, the most meaningful years of my life. Sometimes it has been lonely but I have found my highest fulfilment here among my beloved comrades.'

His devoted translator wrote: 'Our doctor died on the night of many stars in the sky. The heavens wept.'

When Mao Zedong learned of his death, he wrote a eulogy called 'In Memory of Norman Bethune' in which he called him a martyr who selflessly supported the cause of the Chinese people's liberation. Mao urged every Communist to learn from him. During the Cultural Revolution in the 1960s the eulogy was made compulsory reading in every school in China. Millions learned to recite it from memory.

Norman Bethune is buried in the city of Shijiazhuang and his grave has become a place of pilgrimage. Nearby is the Norman Bethune International Peace Hospital and a museum. It contains a glass case displaying his typewriter, some medical instruments and papers – the doctor's only possessions to survive the war.

MEDICINE

DOCTORS LIKE THE innovative Bethune and other men and women who risk their lives to care for the wounded in wartime are universally admired for their courage and dedication. When people flee bombs and bullets in war zones in Africa, Asia and the Middle East, doctors and nurses are often the few who remain to tend the wounded. And modern medicines and surgical methods mean that more survive than ever before. But it was not always like this. Two centuries ago doctors were no less brave but their craft was much more primitive. If a musket ball lodged in your arm or leg in the Napoleonic Wars of the early nineteenth century, you would be lucky to keep the affected limb: the advice would usually be that to avoid infection you should have it amputated. And there would be no anaesthetic to drown the pain. Chloroform wasn't invented till about forty years later. Many stalwart men survived the operations. When Lord Uxbridge had his shattered leg removed after the Battle of Waterloo in 1815, his admiring doctor remarked that his pulse remained the same throughout the agonising operation. He was fortunate to survive. We were so moved by the diary that Magdalene de Lancey wrote about how she nursed her mortally wounded husband after Waterloo that we've chosen her account to begin this chapter.

But people didn't die in the makeshift hospitals of that time only from gunshot wounds. More people died of disease than wounds in the Napoleonic Wars and even as late as the American Civil War in 1861–5 sickness sent more to their graves than gunfire. It would be more than sixty years before penicillin was discovered. By the time of the two great wars of the twentieth century improvements in medicine hugely increased the chances of survival. And by then the doctors struggling to save lives were accompanied by nurses who did much to lighten their load. When casualties mounted some nurses found themselves doing the work of doctors. One, Augusta Chiwy, provided invaluable help to doctors at the Battle of the Bulge in the Ardennes in 1944. She found herself – entirely by chance – in the middle of one of the biggest and bloodiest battles of the Second World War and emerged a Belgian heroine.

Our final choice in this chapter on medicine in war is an American doctor who became renowned not just for his medicine but also for the help he gave to Allied fugitives in Nazi-occupied Paris. Dr Sumner Jackson risked all to rescue Allied airmen downed in France, and somehow managed to fool the Nazi occupiers until, tragically, he himself became their victim.

18

Magdalene de Lancey

Nursing a Dying Husband, Waterloo, 1815

IN MID-OCTOBER 1825, ten years after the great Battle of Waterloo in which the Duke of Wellington finally brought down the French Emperor Napoleon, the great Scottish novelist Sir Walter Scott found himself close to tears as he read a handwritten journal. It was by a young woman who'd lost her husband in the battle. Sir Walter called it 'a heartrending diary. I never read anything which affected my own feelings more strongly.' The account by Magdalene de Lancey describes how William de Lancey, the man she'd married just weeks earlier, died in agony before her eyes of the wounds he'd suffered in the battle. She'd struggled for days to nurse him back to health. She tells the story of how she turned from young bride to devoted carer in touching detail.

Sir William de Lancey came from an old Huguenot family that escaped from France when the Protestants were persecuted in the seventeenth century. He joined the army at the tender age of fourteen and was commissioned as a cornet in the light dragoons. He was dark and handsome with a particularly engaging manner.

Everyone liked him. When he met Arthur Wellesley, the future Duke of Wellington, he was a sixteen-year-old captain. He went on to serve Wellington with distinction in his series of victories over Napoleon's armies in the Peninsular War. Peace came to Europe in 1814 with the defeat and exile of Napoleon to Elba. And on 4 April 1815, when he was thirty-seven and a lieutenant colonel, Sir William married his sweetheart Magdalene Hall, the daughter of a famous Scottish geologist. She was just twenty-two, vivacious, dark-haired and beautiful. 'I cannot,' she writes, 'recollect a day of my short marriage that was not perfect.'

Very soon after the wedding William de Lancey was summoned by Wellington to join him in Brussels. The French emperor had shocked the world by escaping from Elba. He had reassembled what was left of his old Grande Armée and was now determined to destroy the allied armies that were bent on eliminating him once and for all. Britain's allies, Prussia, Russia and Austria, entrusted Wellington with the task of stopping Napoleon from seizing Brussels. And he asked de Lancey, for whom he had the highest esteem, to take a senior role on his staff.

On 9 May, five weeks after his marriage to Magdalene, Sir William de Lancey moved to Brussels. Wellington needed the best men he could get to resist Napoleon and de Lancey would be assistant quartermaster general – responsible for the transmission of orders to the various army units. Like his close friend Harry Smith and many other officers, de Lancey was joined by his wife, who shared his elegant rooms overlooking a leafy square. 'It was amusing enough, sometimes, to see from our windows the people parading in the Park.' William only went to his Brussels office for an hour or so a day. The couple spent every possible moment together. When others dined in the afternoon, they went for a quiet walk. At six they'd have a few friends to dinner. 'I never passed such a delightful time,' she writes, 'for there was always enough of very pleasant society to keep us gay and merry, and the rest of the day was spent in peaceful happiness.' Magdalene determined not to spoil the joy of being with her husband by thinking of the dangers that lay ahead. 'We enjoyed each hour as it passed with no more anxiety than was sufficient to render time precious.'

Then came Thursday, 15 June. Her husband dined with the Spanish ambassador: it was the first invitation he'd accepted since Magdalene arrived in Brussels. 'He was unwilling to go and delayed and still delayed, till at last when near six, I fastened all his medals and crosses on his coat, helped him to put it on and he went.' She watched at the window until he was out of sight and then, 'I continued musing on my happy fate. I saw my husband loved and respected by everyone, my life gliding on, like a gay dream, in his care.'

Little did she know that the storm was about to break and how her happiness would turn to despair. That very afternoon Wellington heard that Napoleon had thrust through the Belgian frontier and attacked the forces of Marshal Blücher, Britain's Prussian ally. The three-day countdown to Waterloo had begun. An hour after William left for his dinner an aide-de-camp rode up and asked Magdalene where he was. She told him and minutes later saw her husband gallop past her to the duke's house a few doors down the street. 'I must confess my courage failed me now, and the succeeding two hours formed a contrast to the happy forenoon.'

Magdalene de Lancey then records how at 9 p.m. her husband came in and saw how unhappy she was. He told her not to worry, that there would be a big battle the next day and that she should go to Antwerp for her own safety. He would probably join her there. He said he would be working and writing all night as he had to get the whole army on the move. He asked her to be ready with some tea when he returned later. She realised he did not want her to ask him any questions and she stayed silent. 'I moved . . . like one stupefied.' He didn't get back until midnight, 'much fatigued, but he did not attempt to sleep; he went twice to the Duke's.'

The duke himself was busily talking to senior officers and deploying his forces, and at the same time dressing for a ball hosted by the Duchess of Richmond. In spite of the sudden approach of the French he'd told the duchess to go ahead with the ball to avoid Brussels being seized by panic. When de Lancey called at the duke's at around 2 a.m., Wellington was back from the ball and fast asleep. He had ordered his men to march south and was now grabbing some rest. At 3 a.m. the troops were all assembled and Magdalene recalled: 'Sir William and I leant over the window and the scene was very solemn and melancholy.' Fifes played as they marched by and 'I saw them melt away through the great gate at the end of the square. Shall I ever forget the tunes played by the shrill fifes and the bugle horns which disturbed that night!'

Early the next morning, the 16th, Magdalene left Brussels and travelled to Antwerp. That evening she was overjoyed to receive a letter from her husband saying 'they had given the French a tremendous beating' but the fighting wasn't over. Wellington had managed to rush enough troops to Quatre Bras, some twenty miles south of Brussels, to stop the French breaking through. But the Prussians had been battered and had pulled back. Wellington withdrew part of the way back to Brussels and formed a defensive line on a ridge just south of a village called Waterloo.

The great battle of that name was fought on 18 June and Magdalene received the first news of Wellington's decisive victory the following morning at 9 a.m. Initially she was told that Sir William de Lancey's name was not on the list of dead or wounded and was thrilled to hear that her husband was safe. 'I was almost in a fever with happiness.' But her joy was short-lived. When Magdalene met the woman who had compiled the list, she was told that in order to spare her feelings de Lancey's name had been left off it. He was in fact seriously wounded.

Frantically Magdalene called for her carriage and had to wait two hours for it to arrive. On the road to Waterloo there was chaos. 'We were entangled in a crowd of waggons, carts, horses, wounded men, deserters or runaways, and all the rabble and confusion . . .'

Halfway from Antwerp, Magdalene spotted an acquaintance, William Hay, who'd been near the front at Waterloo and asked him if he knew anything.

'I fear I have very bad news for you.'

'Tell me at once, is he dead?'

'It is all over,' replied Hay.

Distraught, Magdalene sank back into the carriage and returned to Antwerp where she insisted on being left alone. 'I wandered about the room incessantly, beseeching for mercy, though I felt that now, even Heaven could not be merciful.'

In the early hours of the following morning she heard someone trying to get into her room and she sent her maid Emma to find out who it was. Emma opened the door to a messenger from the battlefield and returned saying, 'I am desired to tell you cautiously—'

'How can you be so inhuman!' interrupted Magdalene. 'What is good news for me now?'

'But,' said Emma, 'Sir William is not dead.'

Magdalene didn't know what to believe. She was again desperate to get to Waterloo. She wouldn't even allow Emma to waste time packing her clothes. 'My agitation and anxiety increased. I had the dreadful idea haunting me that I should arrive perhaps half an hour too late. This got the better of me, and I paced backward and forward in the parlour very fast, and my breathing was like screaming.'

The road was only a little less crowded than it had been earlier, and it was several hours before she arrived at the cottage where Sir William de Lancey had been taken. Yes, said the senior British officer who greeted her, he is alive and the surgeons are of the opinion that he may recover. But he added: 'I wish to warn you of one thing: you must be aware that his life hangs on a very slender thread.'

As she approached his room she heard her husband say, 'Let her come in then.' In an instant her misery vanished. She was surprised how strong de Lancey's voice sounded and when she saw him she couldn't speak. She just took his hand. He asked her if she was a good nurse and she replied that she hadn't much experience of it. 'He said he was sure he would be a good patient, for he would do anything I bade him till he was convalescent. All his endeavour seemed to be to leave none but pleasing impressions on my mind, and as he grew worse and suffered more, his smile was more sweet and his thanks more fervent for everything that was done for him.'

Magdalene learned what had happened at Waterloo. At around three o'clock on the 18th in the middle of the fighting, while de Lancey was riding at Wellington's side, a cannon ball struck him on the back at his right shoulder and knocked him off his horse. The duke thought he must have been killed but then, seeing he was alive, he ran to him, stooped down and took his hand. Eight of his ribs were broken: one had pierced his lung. De Lancey begged the duke to leave him to die in peace, but his cousin Lieutenant Colonel Delancey Barclay insisted on moving him to a cottage in Waterloo. He was left there for the best part of twenty-four hours without attention. The duke assumed he was dead and said so in his Waterloo dispatch.

It wasn't until the following morning that he was found. Once Magdalene was with him the surgeons told her there was a chance that he would survive as long as they could bleed him constantly to reduce the inflammation in his chest. Fortunately the family in the cottage were very friendly and helpful but the bed was primitive and uncomfortable and the pillow so full of dust that Magdalene had to replace it with a cushion from her carriage. 'I had the delight of hearing him say that he did not know what he would have done without me. He said he was sure he would not have lived so long, for he would not have been so obedient to anyone else.' He told his wife that even if he recovered completely, he would never think of serving again: 'We should settle down quietly at home for the rest of our lives,' he said.

Four days after he was wounded, and in spite of Magdalene's constant nursing, Sir William was very feverish, and his blood appeared to be 'very inflammatory'. About ten that night, the doctors returned. 'While I told them how Sir William had been since their last visit, and mentioned several circumstances that had occurred, I watched them and saw they looked at each other. I guessed their thoughts. I turned away to the window and wept.'

Later that night another doctor called, the one who had initially predicted William would recover.

> He was grieved indeed to think that it should fall to his lot to tell me that it was the opinion of the surgeons that if I had anything particular to say to Sir William, I should not delay long. I asked, 'How long?' He said they could not exactly tell. I said, 'Days or hours?' He answered that the present symptoms would certainly not prove fatal within twelve hours. I left him, and went softly into my husband's room, for he was sleeping. I sat down at the other end of the room, and continued looking at him, quite stupefied; I could scarcely see. My mouth was so parched that when I touched it, it felt as dry as the back of my hand. I thought I was to die first. I then thought, what would he do for want of me during the remaining few hours he had to live.

Magdalene believed she should tell her husband how close he was to death.

> I sat thinking about it, when he awoke and held out his hand for me to take my usual station by his bedside. I went and told him. We talked some time on the subject. He was not agitated, but his voice faltered a little, and he said it was sudden. He felt at peace with all the world; he knew he was going to a better one, etc., etc. He repeated most of what he had told me were his feelings before – that he had no sorrow but to part from his wife, no regret but leaving her in misery.

The doctors suggested she try heating a large piece of cloth and put it on his breast to soothe him. There was no flannel available so she tore a piece out of her petticoat, heated it and laid it on his chest. 'He said it did him good and was delicious, unconscious of where we had found the flannel.'

By Friday her sleepless nights sitting by her husband were catching up with her and she asked if she could get some sleep. But he asked her to stay and apply leeches to his body in the hope that more blood loss might relieve him. 'This was the only time I hesitated to oblige him, for I really could scarcely stand, but of course I proceeded to apply the leeches, and in a few minutes the excessive drowsiness went off.' Sir William told her he wished there were some device to shorten the weary long night. 'He said if I could lie beside him it would cut off five or six hours. I said it was impossible, for I was afraid to hurt him, there was so little room.' But Sir William was so beseeching that in the end she agreed and stood on a chair to step over him and lie on the inside. They both managed to get some sleep.

The next morning he had some breakfast and then said he wanted to get up. 'I persuaded him not to do it; he said he would not do anything I was averse to, and he said, "See what control your poor husband is under." He smiled, and drew me so close to him that he could touch my face, and he continued stroking it with his hand for some time.' About three o'clock the doctors returned – looking much graver. William asked to have his wound dressed and while the doctors left the room to prepare a dressing, 'I sat down by my husband and took his hand; he said he wished I would not look so unhappy. I wept; and he spoke to me with so much affection. He repeated every endearing expression. He bade me kiss him. He called me his dear wife.' When the doctors came back in, Magdalene left the room: 'I could not bear to see him suffering.' But moments later she was called back to Sir William's bedside.

'I hastened to him, reproaching myself for having been absent a moment. I stood near my husband, and he looked up at me and said, "Magdalene, my love, the spirits."' She stooped down close to him and held the bottle of lavender to him and also sprinkled some. 'He looked pleased. He gave a little gulp, as if something was in his throat. The doctor said, "Ah, poor De Lancey! He is gone." I pressed my lips to his, and left the room.' Magdalene had nursed her dying husband for eight days since the Battle of Waterloo.

She waited until everyone had left the bedroom and then went back in to spend some time beside William's lifeless body. 'There was such perfect peace and placid calm sweetness in his countenance, that I envied him not a little. He was released: I was left to suffer. I then thought I should not suffer long. As I bent over him I felt as if violent grief would disturb his tranquil rest.'

William de Lancey was buried just outside Brussels. Magdalene had a stone placed on his grave which simply named him and said: 'He was wounded at the Battle of Belle Alliance (Waterloo) on the 18th June 1815.' 'Since that time,' she writes, as she brings her Waterloo diary to a close, 'I have suffered every shade of sorrow, but . . . I have never felt that my lot was unbearable. I do not forget the perfection of my happiness while it lasted; and I believe there are many who after a long life cannot say they have felt so much of it.'

We are not alone in finding this a most affecting story. After reading Magdalene's diary, Charles Dickens wrote: 'I shall never forget the lightest word of it . . . I cannot throw the impression aside, and never saw anything so real, so touching, and so actually present before my eyes.'

Magdalene married again, but died aged twenty-nine giving birth to her third child.

19

Sumner Jackson

An American in Paris, 1940–4

A MONTH BEFORE THE Allied landings on D-Day in 1944, when Hitler's armies still occupied much of Europe, three policemen knocked on the door of an American family's flat in Paris. The son, Phillip, opened the door and the Vichy French police, ordered in by the Gestapo, swept past him and told his mother, Toquette Jackson, that they had come for her husband, Dr Sumner Jackson. She said he was at his surgery in the American Hospital and two of the policemen left immediately to arrest him. Phillip, aged sixteen, walked casually to a front window and rearranged the curtains. It was a signal to the French Resistance that the apartment was no longer safe. The next day all three Jacksons were transported to Nazi extermination camps. It was the horrific climax to lives spent in constant peril in German-occupied Paris.

When Hitler's army seized Paris in June 1940, most sensible Americans had already left. The Second World War had started nine months earlier and the US government had advised its citizens to leave when Germany invaded Poland. Many Americans held France and Paris in particular affection and there had been a long friendship between the two countries since France took the side of the thirteen British colonies that rebelled and declared their independence in 1776. It was a love affair that lasted. The lure of Paris in the 1920s and 1930s meant that 30,000 Americans had moved there including the exotic showgirl Josephine Baker who famously sang that she had two loves, America and Paris. Baker was one of the few Americans brave enough to stay on in France once war began and she soon offered her services to the Resistance.

Another of those who chose to remain was American surgeon Dr Sumner Jackson. He was over six feet tall and powerfully built, with dazzling blue eyes and heavy dark eyebrows. One of his French woman admirers said he looked 'like one of the heroes from a Fenimore Cooper novel'. He had been born in Maine fifty-four years earlier and received his medical training in Philadelphia, where he was described as 'a fine student of excellent character'. During the First World War he worked in France as a doctor on the Somme. His experience coping with the challenges of emergency medicine at the front made him a natural choice as a doctor with the American Red Cross when the United States joined the war in 1917. He married a lively French woman, Toquette, that year and in 1921 the couple moved permanently to Paris when he was invited to join the American Hospital, which offered free treatment to Americans who couldn't afford to pay. The writer Ernest Hemingway became a patient when a skylight in his bathroom fell on his head. Sumner Jackson deftly stitched him up.

With the German invasion of France in May 1940, Sumner Jackson's hospital was overwhelmed with patients. It immediately offered to take in French soldiers wounded in the defence of France. Jackson and his colleagues were engaged in almost non-stop work. He gave his own blood a number of times. Once he had to amputate a French soldier's leg in the darkness after the wounded man's ambulance caught fire. The day before the Germans entered Paris, one of Jackson's American colleagues who was heading back to the States warned him that Americans would be locked up under the Nazis. The USA was not yet at war with Germany but it made no secret of its sympathy with Britain; however, encouraged by Toquette,

Jackson opted to stay. He saw it as his mission to protect his patients and save the hospital from falling into German hands.

It wasn't long before Sumner Jackson set himself a far more dangerous mission. Over the next four years, even after America entered the war against Germany in December 1941, Jackson's hospital became a sanctuary for American and British fugitives. The German occupation regime kept a close eye on Americans but Sumner had influential supporters in the Vichy French community who sided with the Germans. The hospital remained independent enough for Jackson to use it as a haven for opponents of the Nazis. If they were lucky, British and American airmen who were shot down and managed to escape arrest found their way there and Jackson arranged for their onward travel out through Spain or Switzerland. He constantly urged his colleagues in the hospital to keep German soldiers out at all costs. 'No empty beds,' he ordered, 'never a room for the Boches.' Astonishingly most of Jackson's activities remained secret – even to his family.

Young Phillip Jackson knew almost nothing of what his father was up to. Eighty-seven years old when we talked to him in his Paris apartment in 2015, he told us: 'It was a very delicate situation. Any mention of any specific would have endangered what my father was doing. He never told me anything in detail.' One cheeky prank that Phillip picked up from his fellow schoolmates horrified his parents. He would take a piece of chalk from the classroom and chalk up 'V' for victory signs in public places when no one was looking. When Sumner and Toquette found out, they were appalled and told him never to do it again. 'My father was indeed very cross with me for endangering the family,' Phillip told us.

It was a nerve-racking time for the Jacksons. They had a house outside Paris in the quiet lakeside suburb of Enghien, but their Paris apartment was in the smart Avenue Foch just a few doors down from Gestapo headquarters. The French called it Avenue 'Boche'. The Jacksons also found themselves uncomfortably close to another feared Nazi organisation, the SD, the Sicherheitsdienst, the Nazi Party's security agency. Its head Karl Oberg was sent to Paris by Hitler to crush the French Resistance and to round up all Jews. Oberg soon built up a network of French informers to catch Resistance fighters. His victims were usually executed in the Bois de Boulogne.

Oberg's Nazis waited nine months after the United States entered the war before they staged a mass internment of Americans in Paris. Jackson was examining a patient in the hospital when soldiers ordered him to go with them immediately. He remembered hospital staff showering him with gifts as he left. 'We all had tags hung around our necks. That evening we were at [the internment camp at] Compiegne. Nearby was a camp where Jews were held.' Jackson was lucky: this first internment by the Germans didn't last long and he was soon back at work. He made a point of arranging for Jews who were seriously ill to be admitted to his hospital, where he now had the title of chief physician.

Phillip Jackson had one brief glimpse into his father's secret world after he saw German warplanes attack and shoot down an American B-17 bomber over Paris in 1943. The tail-gunner from the B-17 escaped the crash and was smuggled into the Jacksons' flat in Avenue Foch. Phillip caught sight of the airman Joe Manos before he was sent off to the family house at Enghien. He was from New York City and he'd parachuted into a field of sugar beet at Le Bourget. Two Frenchmen packed him into their car, covered him with firewood and eventually he was passed on to Dr Jackson in the hope that he could arrange an escape through Spain. It was a huge risk since the Gestapo just down the road were searching everywhere for Manos and the rest of his surviving crewmen. Sumner Jackson ferried the airman from the hospital to his flat on the back of his bicycle. A friend of the Jacksons in the Resistance forged papers for him and he was successfully evacuated through Spain to Gibraltar and freedom.

A few weeks later the Jacksons had an exceedingly risky request from a leading member of the Resistance who had already persuaded Sumner and Toquette to allow their flat to be used as a safe house in Paris for deliveries and contacts. Now the Resistance asked the couple to allow fifteen-year-old Phillip to carry out a very dangerous mission. The British needed to know exactly what damage Allied bombers had done to the huge area of German U-boat submarine pens at the French port of St Nazaire. The only way to be sure was to have someone take pictures on the ground. But how to get a camera into the forbidden zone around the docks? A fifteen-year-old boy might get away with it. Would Phillip be prepared to try his hand at a vital piece of espionage? Remarkably Sumner and Toquette, who had up to then been very protective of Phillip, agreed. They had friends near St Nazaire and Phillip knew the family's young daughter Rosie. They would make an innocent enough pair of young teenagers exploring the sights of St Nazaire.

Phillip was introduced to a member of the Resistance in St Nazaire codenamed 'Dorsal', and given a simple box Brownie camera. He then went to meet his family's friends, Rosie's parents, Monsieur and Madame Bagousse. But then he made a foolish mistake: he unpacked the camera. M Bagousse immediately confiscated it, saying it was a dangerous thing to be seen carrying. The next day Phillip had the frustrating experience of climbing the bell tower of a deserted church with Rosie, enjoying a magnificent view of the docks but unable to take any snaps. 'If only I had the camera,' he remembered saying. That night he climbed into bed utterly mortified that he'd failed in his mission. But in the early hours there was a little tap on his bedroom door and in walked Rosie, holding the camera. Putting a finger to her mouth she whispered that she'd found it in the linen closet. Phillip's mission was back on.

The next day the two of them were out in St Nazaire again, climbing the church tower and photographing freely. Phillip's film eventually reached London where it was of great value in helping the RAF and the USAF assess the damage done by months of bombing of one of Hitler's prime submarine bases.

By the end of 1943 the war was beginning to turn in the Allies' favour and plans were being laid for the invasion of France. Sumner Jackson and Toquette worked even harder to help the Resistance, unlike many of their French contemporaries who were either collaborating with the Nazis or just keeping silent. In a massive coup for British intelligence a German security officer who had access to the plans of the V1 rockets, which were causing terrible damage around London, defected. He had acquired drawings of the rockets themselves and even the location of the V1 launch sites near the Channel coast. It was decided that Jackson's American Hospital would be the best place for the German to deliver his material. The vital details eventually reached London and Winston Churchill himself praised the men and women who had formed the chain of intelligence on the V1 launch sites. By attacking the sites, he later said, Allied bombers were able to 'mitigate the violence' of the V1. 'To all our sources, many of whom worked among deadly danger, and some of whom will be for ever unknown to us, I pay tribute.'

It was all too good to last. General Oberg's agents had been busy and in May 1944 the Resistance group working with the Jacksons was betrayed and exposed. The Jacksons were taken to a prison in Vichy in central France and locked in different cells. They were at this stage not badly treated or tortured, but Toquette wrote to her sister how desperately she needed to wash. She hadn't undressed since the day of their arrest five days earlier. 'My courage is being tested to the extreme not so much for me as for Peter [her pet name for Phillip] and also for Jack [her husband Jackson]; if I knew that he was free my particular fate would be less painful.' Phillip was in a cell with three other men who were whipped and tortured during questioning. 'I was also whipped by jailers,' he told us, 'if I was not standing to attention.'

In mid-July Sumner Jackson and his son saw Toquette one last time before they were moved to another prison in France. The two men were taken off by bus. There was one moment on the journey when they were allowed to urinate in a field and briefly thought of making a run for it. But since they were handcuffed together Jackson told his son they hadn't a chance. They were packed into cattle wagons and taken by rail to Neuengamme prison on the Elbe. 'At the station we were taken over by the SS, whips in hand and holding their enormous fierce dogs,' recalled Phillip. 'We were horse-whipped out of our carriages by the SS and marched to the camp in batches of 100.' He described being taken to a washhouse where everything they possessed even rings and watches were taken off them. 'Our heads and the remainder of our bodies were shaved. Then we had a bath, after which a most minute search was made of our naked bodies with lamps.'

Toquette was transported to the infamous Ravensbrück prison where she was lucky enough to fall ill and receive special medical treatment. Otherwise she would almost certainly have been shot or sent to the prison's gas chambers. In Neuengamme Phillip Jackson calculated that 35,000 prisoners were killed or died while he and his father were there from mid-July 1944 to 21 April 1945. Conditions were appalling and the pair only survived because they had useful jobs, Phillip in the kitchen, Sumner in the prison forge. Unofficially Sumner was also a doctor to other prisoners.

Another inmate, Michel Hollard, who had been head of a French Resistance group, struck up a close relationship with Jackson, whom he greatly respected and admired. 'He was very upright,' Hollard remembered, 'with white hair, strong features and a stern almost hard expression . . . and he was tall, he stood up straight as an "I".' Hollard met Phillip and Sumner Jackson again when they were unloaded from cattle wagons at the north German port of Lübeck on 21 April 1945. They'd been shifted there from Neuengamme to be held in a prison ship. As they stood on the quay Hollard signalled to Sumner that he and a group of French-speaking prisoners were being evacuated to Sweden. Would he like to join them? Sumner shook his head, pointing to the other prisoners lying sick in the wagon – including his own son – and making it clear that they were his patients and he had to stay with them. Hollard then watched the Jacksons being herded on to the prison ship SS *Thielbek*. By coincidence Toquette was evacuated from Lübeck to Sweden that day and may even have spotted from a distance the file of prisoners from the cattle wagons boarding the *Thielbek*. Her husband and son together with 2,500 other prisoners were crammed into the ship's hold. The Germans planned to sink her and three other ships together with all the prisoners aboard.

On the afternoon of 3 May – just twenty-four hours before German forces surrendered – four RAF bomber squadrons took off to attack 'enemy naval formations' in the bay of Lübeck. Four Typhoons of 198 Squadron attacked the *Thielbek*, thinking she was full of German troops. Sumner Jackson was in the hold, while Phillip had persuaded a guard to let him get a breath of fresh air on the deck. When the rockets struck Phillip opened a hatch and shouted for his father. 'I waited a few minutes but could not see him. The *Thielbek* began to list to starboard. I stripped but remained on deck for another five minutes to try and see my father to jump together in the sea.' There was no sign of Sumner. Phillip threw himself into the sea and 'I swam like never before.' The Germans sent boats from the shore to rescue their own sailors. Phillip managed to climb aboard one of them before they knew he was a prisoner. Over 1,000 other prisoners were left to drown. When the boat crew realised Phillip was a prisoner they threw him overboard and he had to swim to the beach. His ordeal wasn't over even when he landed: the SS put him and other surviving prisoners against a wall and levelled a machine gun at them. It was only the arrival of a British tank moments later that scattered the German gunners and saved Phillip. He tried desperately to find his father but without success.

Phillip and his mother were reunited in Paris two months after the war. Toquette was in poor health for the rest of her life and died at the American Hospital in 1968. She and Phillip were decorated for their service to the French Resistance. Sumner Jackson was also posthumously decorated in France and the United States. His body was never found.

20

Augusta Chiwy

Forgotten Angel of Bastogne, 1944–5

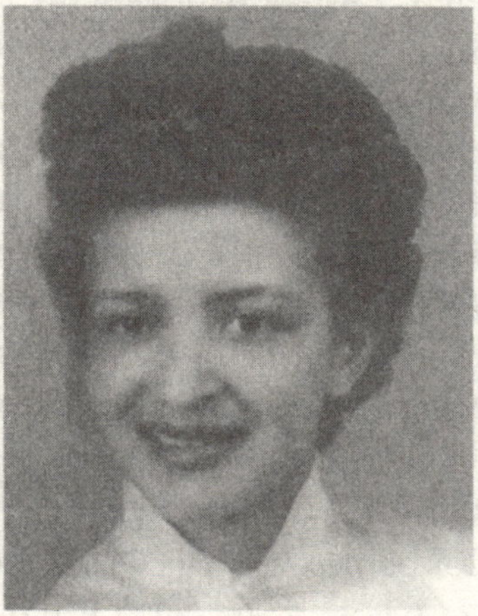

When we watch television together we tend to notice different things. In an enthralling episode of the TV series *Band of Brothers*, Peter was focused on the strategic importance of a key battle in the closing months of the Second World War while Ann noticed an unusual figure among the combatants. Amid battle scenes there are fleeting glimpses of a black nurse tending dying and wounded American soldiers in the small Belgian town of Bastogne. It is almost Christmas 1944 and Hitler has launched a surprise attack on Allied forces – a deadly offensive through the Ardennes forest that will become known as the Battle of the Bulge. The Germans completely surround the town and inflict heavy casualties on the trapped Americans. Within days the dead and wounded run into the hundreds and there are only two nurses and one doctor to cope with a massive medical emergency in appalling conditions. In the TV show the black nurse, referred to only as Anna, is shown assisting a white nurse named Renée. When a US medic asks Renée where 'the black girl' comes from, she replies, 'The Congo.'

We know Renée was based on a real-life heroine. Her last name was Lemaire and she lost her life when, on Christmas Eve, a German plane dropped a bomb on the medical aid station. She became known as the Angel of Bastogne. But who was Anna? It has taken sixty years to solve this mystery. Her name was forgotten but the vital role she played in saving countless lives in those desperate days has now been rediscovered. Thanks to ingenious detective work by a British historian named Martin King we now know her full story. Her name was Augusta Chiwy.

She was born in the Belgian colony of Congo in 1921 where her Belgian father Henri Chiwy worked as a veterinarian. Little is known about her Congolese mother. When Augusta was nine years old her father moved back to Belgium, taking her with him. He asked his sister, a teacher in Bastogne, to help look after her. Augusta wanted a career like her aunt's but at that time black people were not allowed to train as teachers in Belgium. So she became a nurse. During the Second World War, when Belgium was under German occupation, she worked in a hospital near Brussels. The Nazis forced many nurses to work for them but not Chiwy. They didn't like the colour of her skin.

By the end of 1944 the Allied sweep through northern Europe was successfully pushing the Germans back. Allied armies were poised to advance from Belgium into Germany itself. Bastogne had been in American hands for three months so on 16 December Chiwy headed home to spend Christmas with her father and aunt. But at the station in Brussels she was told without explanation that the train would only go as far as Namur, eighty-five kilometres north of Bastogne. Little did Chiwy or her fellow travellers suspect that the train was stopping short of its destination because at that very moment German troops were pouring into Belgium and Luxembourg in a desperate attempt by Hitler to regain territory he'd lost that autumn. It was the beginning of the Ardennes Offensive, the Führer's last great gamble.

Facing ultimate defeat in the face of the Allied advance in the west since D-Day and the collapse of his forces in the east after Stalingrad, Hitler decided to risk all with a massive counterblow. He hoped to exploit reports of disunity between American and British generals by thrusting west and driving a wedge between their two armies. Hitler insider General Alfred Jodl, Chief of Operational Staff of the German high command, discussed the plan, drawing it on a map in Hitler's sickroom when the Führer was ill with jaundice. Jodl later called the counteroffensive an act of desperation. 'By remaining on the defensive, we could not hope to escape the evil fate hanging over us. We had to risk everything.' Some of Hitler's commanders expressed doubts but didn't dare suggest he abandon his plan altogether. As usual Hitler got his way and firmly wrote on the battle orders: 'Not to be altered'.

Hitler chose to attack through the thick forests of the Ardennes because they were thinly held by American troops. The plans to advance along a seventy-five-mile front were drawn up in the utmost secrecy. German troops were not told of their mission until the night before and moved under cover of darkness. The unsuspecting Allies, and civilians like Augusta Chiwy getting ready for Christmas, had no idea what was in store for them. More than one million men on both sides would be involved in this, the biggest and bloodiest single encounter the Americans fought in the Second World War.

On the very day that Chiwy headed home to Bastogne, Nazi tanks made it one of their prime targets as they burst west out of Germany. She only just got there first. On 16 December, after her train terminated at Namur, Chiwy boarded a truck that took her to a spot forty kilometres north of Bastogne. She then managed to hitch a ride with some American soldiers but they couldn't speak French and her English was limited. They drove her too far – to a town twenty kilometres south of Bastogne. She caught a bus and then walked, arriving home just hours before the Germans attacked.

That same evening it became clear to the Supreme Allied Commander General Dwight D. Eisenhower that the Germans had launched a serious offensive, and he ordered all reserve troops immediately into action. On 17 December alone, 60,000 Allied forces, mainly Americans, moved into the Ardennes. Over the next few days a total of 250,000 were deployed.

Allied commanders pinpointed Bastogne as key to any enemy advance. Several highways led in and out of the town: it was an important crossroad hub surrounded by hilly forested country. The US army 8th Corps was already headquartered in the town but not in sufficient numbers to block the Germans. American troops from the 82nd and 101st Airborne Divisions were rushed to the area – some so hurriedly that they didn't have proper maps, winter clothing, rations or even weapons. A soldier from the 101st Airborne remembered carrying just a rifle when he entered Bastogne. Some of his fellow soldiers who didn't have any weapons at all stopped Allied soldiers who were retreating from the Germans and took their guns. As American reserves poured into Bastogne, most of the local population left. But not Augusta Chiwy, her father and aunt. They stayed and took shelter in the cellar of the high school Augusta had attended. As fighting intensified, Chiwy helped nuns and local doctors look after hundreds of local people huddled in the cellars.

Five days after Augusta Chiwy arrived in Bastogne German troops had completely surrounded the town but their repeated savage attacks could not force its surrender. Bastogne's American defenders paid a high price. Not only were they running out of ammunition, food and medical supplies but with each passing hour the death toll rose.

On 21 December Chiwy's father had a visit from Captain Jack Prior, an American doctor with the 10th Armoured Division. He had just arrived in Bastogne and was the only doctor in the aid station crammed with over 100 injured soldiers and civilians. He later wrote: 'The patients who had head, chest and abdominal wounds could only face certain slow death since there was no chance of surgical procedures – we had no surgical talent among us and there was not so much as a can of ether or a scalpel to be had in the city.' Prior had heard that Augusta Chiwy, a trained nurse, was in town and he asked her father if she could help him. Augusta was struck by his kind face: 'he told me that he had no one left, that his ambulance driver had been killed.' She spoke only a few words of English, but Prior knew a little French and she agreed to go with him to his medical aid station located in an abandoned three-storey building. She said she could smell the place before she went in – the stench of disease and death. It was something she would remember for the rest of her life. Chiwy went straight to work injecting a dying soldier with morphine. She remembered Dr Prior standing over her, watching closely. 'Are you looking at me to see if I know what I am doing?' she asked. He was so impressed he left her to work unsupervised from then on.

Dr Prior had also recruited Bastogne's only other trained nurse, Renée Lemaire. The two women performed crucial but very different roles. Prior later wrote: 'Renée shrank away from the fresh, gory trauma' preferring to 'circulate among the litter patients, spongeing, feeding them and distributing the few medications we had (sulfa pills and plasma)'. Meanwhile Augusta was always in the thick of things. After her clothing became bloodstained, she put on a US military uniform. She later said she was very nervous about working for the Americans 'because some of them didn't want a black nurse touching them'. When wounded soldiers objected to being treated 'by a black girl', Dr Prior gave them a simple choice: 'You either let her treat you or you die.'

On her very first day of work, Prior woke Chiwy at four in the morning to say he had found gangrene in wounds on a soldier's hand and foot. After giving the soldier a slug of cognac, the doctor removed the infected limbs with his serrated army knife while Chiwy held the patient down. As soon as Prior finished his amputations, Chiwy cleaned the wounds with hydrogen peroxide, stitched them up and wrapped them in bandages. Jeff Prior, the doctor's son, told us how his father described Chiwy 'playing the role of a pseudo-doctor, splinting broken bones, putting on bandages and doing her best to prevent blood loss'. She also found time to set up a Christmas tree to cheer up her patients. There were no decorations to be had so she decked the boughs with tins from soldiers' ration boxes.

As the fighting intensified, so did Chiwy's workload. As well as nursing the seriously wounded, she treated fast-rising numbers of cases of frostbite and trench foot. The weather was unusually cold, the frozen ground covered with deep snow. Conditions for American soldiers who had rushed to Bastogne without proper winter clothing were hellish. They had no long winter underwear or woollen socks, just battle uniforms and trenchcoats to keep them warm. They had to rely on bedsheets and burlap bags to cover their hands and feet. For the first few days of fighting bad visibility dashed any hope of air drops of clothing, equipment or medical supplies.

On 22 December, convinced that the troops defending Bastogne could not hold out much longer, the Germans sent four messengers into town with an ultimatum: surrender or face annihilation.

American Brigadier General Anthony McAuliffe gave his reply in one word: 'Nuts!' German translators puzzled over the word before deciding – correctly – it meant 'Go to hell!' That same day a paratrooper arrived at the medical aid station asking for help. He said several American soldiers had been wounded in a German attack on nearby Mardasson Hill and needed urgent medical help. When Dr Prior jumped into a truck, Chiwy volunteered to go with him.

At the hill a firefight was raging. Prior told Chiwy to stay low as they crawled towards injured men lying in the snow. Chiwy remembered bullets whizzing past her. The first soldier they reached had been killed by a wound to the head. Another had blood gushing from his neck and chest. He was clearly dying. When they came across a soldier who was unconscious, they acted quickly. Prior used a corkscrew to relieve pressure from the back of the man's head while Chiwy stitched up the wound. As they carried their patient to a truck, the Germans spotted them and fired a mortar shell. It just missed. Chiwy spotted another soldier whose leg had been hit by shrapnel. She treated it with sulfa powder to prevent infection and bandaged it up. Then the tiny nurse, just four foot eight inches tall, helped haul the man to safety. While he pushed himself towards the truck with his good leg, she pulled for all she was worth. Chiwy remembered snow and earth flying up into the air around her as they dodged machine-gun fire. Later she discovered the coat she had been wearing was full of holes. When Prior joked that the Germans missed her because she was so small she replied, 'You're wrong. My black face in the white snow should have made me an easy target. The Germans were just poor shots.'

On 23 December the skies above Bastogne cleared and Allied planes were finally able go into action against the Germans. They could also start air drops of much needed ammunition, food, blankets and medicine. When Nurse Renée Lemaire saw parachutes descending she rushed from the first-aid station and tried to pick one up. She was engaged to be married and wanted to use the silk for a wedding dress. But, wrote Dr Prior: 'She invariably was beaten out by a soldier and always returned empty handed.' But he promised to find her a silk chute.

Sadly Lemaire's wedding dream never came true. On 24 December a fellow officer reminded Prior it was Christmas Eve and invited him to share a glass of champagne in the house next to the medical aid centre. Chiwy went along too. They were about to take their first sip of bubbly when, in Prior's words 'the room, which was well blacked out, became as bright as an arc welder's torch.' A German bomber pilot had dropped magnesium flares above Bastogne to light the area for bombardment. 'We hit the floor as a terrible explosion next door rocked our building. I ran outside to discover the three-story apartment building next-door serving as my hospital was a flaming pile of debris about six feet high.' A 500lb bomb had landed directly on the aid station. 'My men and I raced to the top of the debris and began flinging burning timber aside looking for the wounded, some of whom were shrieking for help.' All thirty seriously wounded men on the first floor died instantly along with Nurse Renée Lemaire.

Chiwy was thrown through a glass wall by the explosion but escaped with a few scratches. She joined Prior and the rescue team and they managed to pull a few patients out alive before the building collapsed completely. Some were so badly burned that they begged to be shot. As German bombers strafed them with machine guns from above, the rescuers frantically searched through the debris and found Lemaire's body blown in half. Prior wrapped her in a parachute and carried the body home to her parents. Chiwy was now the only trained nurse he had to help him. She was not a close friend of Lemaire's but was severely shaken by her death. She remembered going to her father's house and spending Christmas Day in the cellar, shaking uncontrollably.

On Boxing Day, 26 December, Dr Prior begged Chiwy to resume her nursing duties. He was now working out of the Heinz Barracks, headquarters of the 101st Airborne Division. More than 700 injured soldiers lay in litters on the floor of an enormous underground hall once used for riding lessons. Chiwy agreed to help Prior but he noticed she was much changed. She worked as hard and as efficiently as ever but she was unable to speak. When she saw a soldier who'd been blown up by a landmine she tried to scream but no sound came out.

Boxing Day was also the day that General Patton's 3rd Army broke through the German lines to the south of Bastogne and ended the punishing siege. Trucks brought in supplies of food and medicine. Lines of ambulances started evacuating the wounded. Fighting continued around the town for another three weeks until the Germans finally retreated, leaving both sides to count the horrific cost of the Battle of the Bulge: 75,000 Americans and 80,000 Germans dead, wounded or captured. On 17 January 1945, a month after Augusta Chiwy had returned home for Christmas, the 10th Armoured Division left Bastogne. Chiwy's father begged Dr Prior to take her with him. He thought she would be a lot safer with US army medics than staying behind in Bastogne. Prior later wrote that Chiwy's father was 'most distraught with me for refusing to take her along'. But the Germans never returned and four months later the war was over.

Chiwy didn't go back to nursing full time for another twenty years. She married a Belgian soldier and had two children. She never talked to them about the war or her role in it. But she did stay in touch with Dr Jack Prior. She sent him a birthday card each year and they exchanged Christmas letters and gifts of chocolates. 'My dad got the best deal in that exchange,' Jeff Prior told us. 'He got the Belgian chocolates. Augusta got the American stuff.' Jack Prior ended each of his Christmas letters with the words 'I still think about Christmas Eve 1944.'

It took historian Martin King's dedicated search for Chiwy to bring her story to light. After various false starts he managed to find her in 2009. He told us 'she was almost abandoned in a geriatric retirement home in a Brussels suburb.' Determined that Chiwy should be honoured for her war work King wrote dozens of letters in her support. In 2011, sixty-seven years after the Battle of the Bulge, she was made a Knight of the Order of the Crown by King Albert II of Belgium. That same year she received the US army's civilian award for humanitarian service. The colonel presenting the award said she embodied what was best and most kind in human beings. Chiwy modestly replied: 'What I did was normal. I would have done it for anyone. We are all children of God.' In 2014 Chiwy was named a Bastogne Citizen of Honour. She died a year later at the age of ninety-four.

WAR AT SEA

WE ARE BOTH sailors and have a healthy respect for the sea, which can suddenly throw up unexpected perils and challenges. So we felt our book would not be complete without telling the stories of brave mariners.

Most great empires have been built on command of the sea. Athens controlled the Aegean Sea at the height of its power in the fifth century BC. Rome was master of the Mediterranean. China's greatest dynasties made their power felt in the Indian Ocean. Britain, the Netherlands and France fought for control of the oceans in the seventeenth and eighteenth centuries and Britain ended up ruling the waves. The destruction of the French and Spanish fleets at Trafalgar in 1805 allowed Britain's Royal Navy to blockade and finally destroy Napoleon and to extend the British Empire into all the world's continents in the nineteenth century. The twentieth century saw the introduction of giant dreadnoughts before the First World War and then their eclipse by the lethal power and range of the submarine (the U-boat in Germany) and the aircraft carrier. Carrier-based warplanes could far outrange the guns of any great battleship, and one of the greatest combats between aircraft carriers led to America's victory over Japan at the Battle of Midway in June 1942. Germany's U-boats came near to cutting Britain's maritime supply lines in the early part of the Second World War.

We've chosen to illustrate the way the submarine shot up the table of naval firepower with Martin Nasmith's unparalleled voyage through the Turkish defences in the Dardanelles to cause havoc at the heart of the Ottoman Empire – in Istanbul itself – in 1915.

We are also intrigued by the story of Herbert Lightoller, who survived the disastrous sinking of the *Titanic* in 1912 and then, twenty-eight years later, went on to brave German bombardment and take his small motor launch across to Dunkirk in 1940. How he managed to reach the beaches and rescue so many beleaguered British troops is an awesome tale.

But we begin 200 years earlier: we've always found it hard to imagine how early sailors faced the dangers of the sea without modern technology – without engines, radios or modern navigational aids. One of the most extraordinary exploits of those primitive days was how John Bulkeley and the crew of HMS *Wager* survived their shipwreck in the war with Spain on the wild coast of Chile back in 1741.

21

John Bulkeley

The Wreck of the Wager, *1741*

SIX HUNDRED MILES up the western coast of South America from Cape Horn the Pacific Ocean takes a huge bite out of the coast of Chile. It's the Golfo de Penas – the aptly named Gulf of Sorrows. And on its southern side is a large island called Wager Island marked by a conical peak named Mount Anson. A thousand miles south of Valparaiso, it was one of the most desolate and inhospitable places on earth in the eighteenth century with just a few inhabitants eking out a living by fishing and hunting. More than two and a half centuries ago a British warship in a squadron led by Commodore George Anson was shipwrecked there. The vicious infighting among the survivors that led to mutiny and rival attempts to escape through the world's most murderous seas make this one of the most riveting stories in maritime history.

Anson's assignment was to attack and destroy any Spanish ships or coastal settlements he found west of Cape Horn. In 1741 Britain had been at war with Spain for two years, the two great empires locked in a struggle over trade and colonies. The country now known as Chile was fast becoming a prosperous Spanish colony and the British were challenging the Spanish monopoly in South American trade. Anson had six warships under his command including *Wager* and two supply ships all loaded with stores for a long voyage. *Wager* carried guns and ammunition for land operations and she was stuffed with supplies of wine and brandy and other goods to buy the friendship of local natives. Her decks were also crammed with extra passengers, marines and soldiers – a bizarre collection of raw recruits and veterans – some over seventy years old. Many of them had been press-ganged into serving – some snatched from jail. The packed conditions aboard were not much better than prison: fresh food was in short supply so scurvy was rife and hordes of lice and other vermin caused deaths from typhus and dysentery. Moreover, in rounding Cape Horn the men had to tackle monstrous seas and winds that left the ship almost unmanageable. And, as the southern winter set in, worse was to come. *Wager*'s captain, David Cheap, a brave seaman but volatile and impulsive, had only 152 crewmen surviving out of an original total of 243 as he battled up the west coast of South America.

By the beginning of May 1741 *Wager* had lost sight of the rest of Anson's squadron in the atrocious weather, and by the middle of the month John Bulkeley, an enterprising and capable gunner who kept a daily journal, reported that the ship was 'no better than a wreck; the mizzen-mast gone, the standing rigging and chain plates afore and abaft mostly broke and ruined. The topsails now at the yards are so bad, that if we attempt to loose 'em for making sail, we are in danger of splitting 'em.'

Bulkeley had become one of the most respected members of the crew, and when the captain fell and dislocated his shoulder on 13 May he asked Bulkeley to help the first lieutenant run the ship. It was a terrible night, the wind howling and the sea tossing the vessel about and causing such injury and seasickness that Bulkeley had only three fit men available to man the rigging on his watch. Suddenly the ship struck a rock: the crew struggled to free her but she struck again and was soon trapped between rocks only a few yards from the shore of the island that would later be named Wager. A boat was launched and the lieutenant and the injured captain went ashore with a party to try to make a base there. On the battered ship there was near pandemonium. Several of the men were 'so thoughtless of their danger', reported Bulkeley, 'so stupid, and insensible of their misery, that . . . they fell into the most violent outrage and disorder'. They opened the wine casks, broke open cabins and chests and armed themselves with swords and pistols. Then 'being drunk and mad with liquor' they stole money and clothes and pranced around 'imagining themselves lords paramount'.

Amid all this chaos Bulkeley made it ashore with the ship's carpenter John Cummins and found a place to sleep. They took shelter from the howling wind and crashing sea under a tree near a small hut built by local inhabitants that had been commandeered by the captain.

Most of the crew, including most notably the boatswain, John King, who should have kept order, abandoned all sense of responsibility. They were lost in bitter recrimination and mindless negligence and were furious at the captain for the shipwreck and jealously fighting each other for what they could salvage from the ship. Bulkeley made several visits to the *Wager* to unload stores and each time found men dead aboard. He assumed they had been drunk and unable to survive the constant rush of seawater into the ship. By this time – a week after the shipwreck – there was near anarchy among the crew members with the captain quite unable to impose any discipline. Some stories suggest men ate the flesh of their dead shipmates and that a cabin boy was found eating a human liver. The survivors had broken up into a number of cabals: some remained aboard, a few stuck with the captain and others stayed with Bulkeley and Cummins. Ten men announced they were deserting and prepared to make a raft from the ship's timbers to cross to the nearby mainland. Bulkeley attempted but failed to persuade them to stay. He and Cummins assured the captain that 'the people in our tent were generally well affected to him and that we would never engage in any mutiny against him.' But Captain Cheap soon made things worse: hearing that one of his midshipmen was about to mutiny he impetuously shot the man at point-blank range, mortally wounding him. With the captain now increasingly unpopular and men dying of exposure and starvation, Bulkeley and others began to think desperately of ways to escape from this godforsaken place with its unremitting rain and bitter wind.

By August, amid all the misery, there were two small rays of hope. The crewmen had managed to build thatched shacks for shelter and John Cummins, the ship's carpenter, had spent weeks shoring up the timbers on the ship's damaged longboat and turned it into a schooner that now looked seaworthy enough to carry a large number of men. Bulkeley decided to try out an ambitious plan on the captain and the other survivors. He suggested heading south to the Straits of Magellan and attempting to sail this small vessel around the bottom of South America through to the Atlantic. From there, they would sail up the east coast of South America to the friendly Portuguese state of Brazil. The alternative, wrote Bulkeley, was to throw themselves 'into the hands of a cruel, barbarous and insulting enemy' (the Spaniards who lived further north on the Chilean coast). At first Captain Cheap and the others appeared to agree. But they had serious reservations and when Bulkeley and his friends took a letter signed by forty-six of the crewmen urging the captain to formally endorse the plan, Cheap demurred. He said it filled him with great uneasiness and he hadn't been able to sleep thinking about it. 'We are above 160 leagues [around 550 miles] to the Straits of Magellan – with the wind against us.' Bulkeley disagreed with his captain's claim, answering, 'Let the navigators work it, and they will find it not above 90 leagues [300 miles].' Bulkeley was actually the nearer to the truth. The distance to the entrance to the straits is 350 miles. For several days the debate went on, the captain favouring heading north and Bulkeley and most others opting for the southerly course.

By 9 October, five months after being shipwrecked, most of the crew members who had survived the battering of the wind and the desperate scramble for food were so exasperated at the captain's indecisiveness that they resolved on mutiny. They would commandeer the longboat, and the much smaller cutter and head south. Anything was better than remaining in that cheerless spot. Bulkeley and a squad of other crewmen burst into Cheap's tent and told him he was under arrest for the murder of the midshipman four months earlier. They told him they would take him as their prisoner to England. Cheap, supported by the ship's surgeon and another officer, said he preferred to be left on the island than to risk heading south, and after four more days of argument and recrimination the crew split in two. Twenty stayed with the captain, determined to head north in two small open boats and take their chances with the Spanish; seventy-three, including Bulkeley and John Cummins, took the small cutter and the refurbished longboat, which they named *Speedwell*, and headed south.

It was the last week in October. Gales and heavy seas and fast dwindling stocks of food soon made the men exhausted and utterly demoralised. The *Speedwell* frequently lost sight of the cutter and on the night of 6 November huge waves swept the smaller boat away; she and the one man aboard her were never seen again. This caused such consternation that fourteen of the remaining crewmen opted to be put ashore. Bulkeley wrote: 'We wanted to know what could induce them to request our putting them ashore in this remote and desolate part of the world; they answered they did not fear doing well.' They hoped to find the cutter or if necessary make themselves a canoe. They landed and disappeared without trace.

On 10 November Bulkeley guessed they were within reach of the opening of the Straits of Magellan. 'I never in my life, in any part of the world, have seen such a sea as runs here; we expected every wave to swallow us and the boat to founder . . . the boldest men among us were dismay'd.' Their store of food had now run out but they fortunately spotted some natives who agreed to swop 'a mangy dog, which they parted with . . . for a pair of cloth trousers . . . this dog was soon kill'd, dress'd and devoured.' A week later Bulkeley recorded *Speedwell*'s first death. 'George Bateman, a boy aged sixteen; this poor creature starv'd, perish'd, and died a skeleton, for want of food. There are several more in the same miserable condition.' Some men still had a few handfuls of flour or slivers of beef, but no one was going to share what they had. 'Hunger is void of all compassion,' remarked Bulkeley, 'every person was so intent on the preservation of his own life that he was regardless of another's.'

On 19 November a twelve-year-old lad starved to death when his guardian refused to sell the boy's silver watch and chain for flour. Eleven days later three more died. From then on Bulkeley's journal recorded an almost daily toll. There was some relief just before Christmas Day when they managed to kill 'more seal in half an hour than we could carry off', but by New Year's Day 1742 the seal meat was stinking horribly because they had no salt to cure it. 'We are now miserable beyond description, having nothing to feed on ourselves and at the same time almost eaten up by vermin.' By 10 January there were only forty-three of the seventy-three still alive and 'not above fifteen of us healthy . . . I am reckoned at present one of the strongest men in the boat, yet can hardly stand on my legs ten minutes together.'

In spite of all this suffering Bulkeley and his mates had now managed to pass way beyond the Magellan Straits and up the coast of what would later become Argentina as far as what is today the town of Mar del Plata. The shore looked most welcoming and was full of wildlife. They took the longboat close in and were in the act of loading it with supplies when a stiff sea breeze got up and they were forced to pull away, leaving eight men on the beach. The boat spent the night at anchor but a gale the next day broke off the head of the rudder and threatened to smash the vessel. In desperation they sent a large cask ashore with some muskets, powder, flints and candles for their stranded colleagues and a message explaining that they could not risk anchoring off that beach and would have to sail on. The last thing Bulkeley said he saw was the men falling on their knees and making signals 'wishing us well'. This was not how one of the stranded men who survived remembered it. Three of them eventually made it home after five years of starvation, slavery and imprisonment by the Spanish, and they described their desertion by Bulkeley and the others as an act of deliberate abandonment.

By 26 January the remaining crew on the *Speedwell* were once again out of food and Bulkeley recorded: 'This day died the oldest man belonging to us, Thomas Maclean, cook, aged 82 years.' Two days later the crew, now reduced to thirty men, sailed *Speedwell* into the port of Rio Grande just inside Brazil. They had travelled some 2,000 miles around the foot of South America. They were well received and the stories they told of their voyage from the Pacific fascinated the local people, who housed them and fed them well.

It was a whole eleven months later that Bulkeley and two others including John Cummins reached Portsmouth on 1 January 1743 – only to find that they were to be detained on board the ship they arrived on until the Lords of the Admiralty decided what to do with them. 'This was of very great affliction to us; and the more so, because we thought our troubles were at an end . . . Our families had long given us over for lost . . . and our relatives looked upon us as sons, husbands and fathers restored to them in a miraculous manner. Our being detain'd on board gave them great anxiety.' They were soon released but they could not be fully absolved until the admirals were able to unravel the truth of what had happened to the *Wager*. Bulkeley did not wait for that. He rushed his journal into print and it became a bestseller.

Imagine the shock to John Bulkeley and the other mutineers when three years later in 1746 Captain Cheap and two other survivors, including John Byron, the grandfather of the poet, arrived home. The story they told of the suffering they had endured in Spanish jails was almost as dramatic as the astonishing feat of the *Speedwell*'s crew. In spite of fears that the Admiralty would deal out severe retribution for the mutiny everyone was finally cleared by a court martial in April 1746 and both Bulkeley and Cheap were offered ships to command. Bulkeley's was the *Royal George*, a seventy-gun warship with a crew of sixty. He took one look at her and refused the offer. She was, he said, 'too small to bear the sea'. He was right. On her next voyage she sank in the Bay of Biscay with the loss of all hands. There is no record of when this doughty seaman died but we like to think it was at a ripe old age and in his bed.

22

Martin Nasmith

Submariner on the Rampage, Gallipoli, 1915

THE SOUND WAS unmistakable. A dull clang as something hit the nose of the submarine and then a metallic rasping all the way along the port side of the hull. None of the crew could see what it was: the submarine was eighty feet below the surface. But the vessel's captain knew very well that it posed a deadly threat. They were scraping past a metal hawser attached at one end to the bottom of the sea and at the other to a large round mine floating just below the surface above them. 'Stop port,' ordered Lieutenant Commander Martin Nasmith. The left-hand propeller at the aft end of the boat stopped dead. If it tangled with the mooring line it would only be seconds before the mine was sucked down on to the submarine. Nasmith knew just what might happen next: the lethal horns on each mine were on its upper side, but in the turbulence that would follow the fouling of its hawser the mine was likely to tilt far enough sideways in the water for one of the horns to strike the hull. Instantly a glass tube inside the horn would shatter and release acid that would complete the electrical circuit and detonate an explosion large enough to sink a battleship. The submarine's crew froze. The only sounds were the eerie scrape of the wire and the hum of the starboard electric motor. For an instant the hawser seemed to catch on the cowl that covered the propeller and then it was gone. Nasmith restarted the port engine, but he had to repeat this heart-stopping operation several times until Submarine E11 was through the minefield.

It was 19 May 1915. Martin Nasmith and his crew were embarking on one of the most perilous naval expeditions of all time. Allied forces had just landed on the Gallipoli peninsula at the entrance to one of the world's most important waterways – between the Mediterranean and the Black Sea. It was Turkey's only maritime opening to the outside world and Britain and its allies were determined to strike a body blow at Germany's main ally on the eastern front. If they could knock Turkey out of the First World War, the pressure on Russia would be eased and Germany would be gravely weakened. But the allies were meeting major resistance at Gallipoli and the navy's task was to seize control of the Dardanelles. When we visited the peninsula in the centenary year of 2015 it was hard to imagine this tranquil spot was the scene of one of the most controversial battles in the First World War.

The Turks were mounting a formidable defence of the Gallipoli peninsula and Turkish warships and mines were strung across the narrow sea passage of the Dardanelles. These treacherous waters – in places only a mile wide – separated the Aegean, which was under allied control, from the Turkish-controlled Sea of Marmara and beyond that the Black Sea. Submarine E11 had the daunting task of passing through the Dardanelles without being seen. She was then expected to attack Turkish warships and other vessels in the Marmara that were supplying ammunition and other vital needs to their troops in Gallipoli. It was a near suicidal operation for Nasmith and his thirty-man crew. But, if successful, the rewards for Britain and its allies would be immense. Winston Churchill, First Lord of the Admiralty, had taken a risky but bold gamble invading the Turkish mainland. The Sea of Marmara, with the great city of Constantinople at its northern end, was the heart of Turkey's Ottoman Empire. A submarine at large in this inland sea could play havoc with Turkey's communications with its army. If Churchill's navy could penetrate through to attack Turkey's rear the long-term prospects for the allies would be immeasurably enhanced.

Martin Nasmith was one of Britain's top sailors. He was powerfully built without being too tall for a submariner, and his knowledge of underwater warfare, now emerging from its infancy in the nineteenth century, was second to none. He had been chief instructor in submarine tactics in Portsmouth and invented the retractable periscope. But so far his war had not been a success. In spite of his renowned efficiency, his actions had been marred by mechanical failure and bad luck. He was determined that this enterprise would not fail. His submarine, an E class launched thirteen months earlier, had no gun on deck but carried ten torpedoes. There were five torpedo tubes, two in front, two at the sides and one at the back. She was powered by two diesel engines that could propel her at fifteen knots on the surface and two electric motors that gave her a submerged speed of ten knots. A submarine was most vulnerable when she was forced to surface every twelve hours or so to breathe and recharge batteries that drove the electric motors.

The Dardanelles presented a formidable challenge. Five allied submarines had tried to penetrate the thirty-eight-mile-long strait. Four of them had been lost. The Dardanelles were well defended: where there were no mines and warships, there were Turkish shore batteries. In addition, strange currents often moved in different directions on the surface and beneath it. Nasmith had no clever modern navigation aids to pinpoint his position. He had to rely on his compass and the view through his periscope when he took the risk of raising it above the water. The trouble was that even at night a deployed periscope created a small white-water wash easily seen by observers on patrolling ships and ashore. None of this deterred Nasmith: the words of his commodore, Roger Keyes, rang in his ears: 'Go and run amuck in the Marmara!'

At 3.50 a.m. on 19 May, as he entered the Dardanelles, Nasmith gave the order: 'Diving stations!' He sounded no klaxons: the shore, dotted with Turkish observers, was only a mile or so away.

E11's second-in-command, Lieutenant D'Oyly Hughes, called back, 'Both engine clutches out. Shut off for diving!'

'Start the [electric] motors,' ordered Nasmith, taking one last look out of the turret at the direction he'd now have to steer. 'Flood the main ballast!' Water poured into the empty ballast compartments. At the same time the hydroplanes, fins on the side of the submarine, were directed downwards and she submerged. 'Take her down to eighty feet,' instructed Nasmith. That would take them well below the mines lurking near the surface but no dive would avoid the mines' metal hawsers attached to the bottom. Almost every hour Nasmith ordered the submarine up to thirty feet and raised the periscope but each time he spotted blinding shore lights that would illuminate his wake if he stayed on the surface. At 5.30 he wrote in the log: 'Sighted hostile battleship to the northward, who with several hostile destroyers opened fire on our periscope.' At 6.10 a.m., when he again risked showing his periscope to check his position, E11 came under heavy fire from the destroyers. His navigator, Lieutenant Martin Brown, made this log entry: 'The conditions of light, together with speed at which the vessel was travelling, rendered the periscope exceptionally visible.'

All this time the crew, who heard the crashing of shell and small arms fire close to the hull, were applying their well-practised drill of avoiding any unnecessary movement. Oxygen had to be conserved below the surface: any exertion of the lungs added to the drain on the submarine's limited supply.

At 6.40 a.m. a quick glance through the periscope picked up two huge Turkish battleships ahead. They might make a target. 'Flood the tubes!' called out Nasmith. A hiss of air forced water into the torpedo tubes.

'Tubes full, sir!' came the shout a few moments later. But the periscope had attracted fire from destroyers nearby and Nasmith had to dive again. The next time he dared break the surface, the Turkish warships were steaming way off in front of him out of his torpedo range.

By 9.30 a.m. E11 was well past the narrowest part of the strait but Nasmith knew he could not risk surfacing until it widened out into the Marmara. The circulating fans were doing their best to spin around the dank heavy air inside the vessel but oxygen was running dangerously low. Nasmith reduced speed to conserve battery power and it wasn't till it was fully dark at 9 p.m. that he could write in his log: 'Rose to the surface . . . Commenced charge.' They had arrived in the Sea of Marmara. The Dardanelles were well behind them. The hatches were opened and fresh air poured in to the immense relief of the crewmen sweltering in the heat and desperate for oxygen. That night they had to interrupt their charging and dive to avoid two Turkish destroyers. The enemy vessels passed by without spotting the submarine but it meant a second night without much sleep.

The next two days gave Nasmith and crew time to fully charge their batteries and even take a swim in the warm waters of the Marmara. They were soon in action again when they spotted a two-masted Turkish sailing vessel carrying a load of timber. They jumped aboard and so terrified the captain that he begged, 'Don't shoot, good English!' and meekly handed over four fat chickens.

The real fight began on 23 May when Nasmith approached the great harbour of Constantinople. He had long had an ambition to take his submarine right into the enemy's heartland. Now was his chance. Anchored outside was a Turkish gunboat: 'Bearing sixty-three degrees,' announced Nasmith's navigator.

Determined not to allow this one to escape, Nasmith ordered the bow torpedo tubes flooded and kept E11 on a heading of sixty-three degrees. 'Bring her up to twenty feet,' ordered Nasmith. 'Steady: we mustn't break the surface.' The small gunboat looked an easy target, motionless and lying across E11's path. 'Standby the starboard bow tube.' Then, his eyes glued to the periscope, he shouted, 'Fire!' The submarine clattered and shuddered as the seventeen-foot-long torpedo hissed out of the tube. Nasmith watched it hit the enemy vessel exactly amidships. He saw a Turkish seaman race along the deck of the doomed gunboat, grab its six-pounder gun, swing it towards the submarine, and fire one shot which missed. A second shot left Nasmith gasping. It was an exceptionally lucky shot: it hit E11's periscope and damaged it so badly that it jammed. Nasmith had to make a hasty underwater retreat from the harbour mouth and didn't surface until he had found a safe spot in a remote part of the Marmara's southern coast.

The broken piece of periscope was removed and the next day

E11 destroyed her first major target. It was a large cargo ship heading west towards the battlefield at Gallipoli. Nasmith surfaced and rifle fire from the submarine soon brought the Turkish vessel to a stop. He ordered the crew off the ship so that his men could lay an explosive charge to blow her up. But then, Nasmith recorded: 'An American gentleman appeared on the upper deck who informed us that his name was Silas Q. Swing of the "Chicago Sun", and that he was pleased to make our acquaintance.' Swing was very polite and appeared to suggest the ship might not be carrying any military cargo, but a search soon found that she was heavily loaded with six-inch shells and large cartridge cases for giant artillery guns. 'Right,' said Nasmith, 'we'll blow her up.' Swing was asked why he hadn't noticed the military stores on board, and he replied with a smile, 'The United States is strictly neutral.' The enthusiastic journalist, who was obviously thrilled with his story, was waved off by Nasmith into the ship's boats along with the Turkish crew. He then ordered demolition charges laid and detonated. The ship sank in seconds. Within a few minutes Nasmith was chasing another steamer which raced into a port and berthed at a quay. Nasmith loosed off a torpedo at her and blew her to pieces. A little later he fired another torpedo at a paddle boat but missed her.

That night Nasmith and his crew studied charts of Constantinople harbour. He was determined to stage what he knew would be a sensational coup – the sinking of a ship in the enemy's capital. He wouldn't only have to dodge mines and gunboats. The city was at the mouth of the Bosporus, the great strait that led to the Black Sea, and currents and cross-currents were known to flow in all directions at this spot. For example, there had been reports of a counter-current twenty feet below the surface which would hamper navigation. But none of that was going to stop Nasmith. At 6 a.m. he sent E11 into a dive and she approached Constantinople under the surface. Just after noon up went the periscope and Nasmith gazed at the glorious panorama of Ottoman architecture and the frenzy of naval activity that surrounded him. For a moment he was able to relish his achievement. He was in the unprecedented position of being the first hostile vessel to be able to claim to have entered the harbour of Constantinople in some 500 years.

Within seconds his position swung from the sublime to the ridiculous. He recalled later:

> Our manoeuvring was rather difficult because of the cross tides, the mud and the current, but most particularly on account of a damn fool of a fisherman who kept trying to grab the top of my periscope every time I raised it to take an observation. I don't think he had any idea what it was, but to get rid of him I gave him a chance to get a good hold of it. Then I ordered 'Down periscope quickly!' and almost succeeded in capsizing his boat.

Things then went from bad to worse. Nasmith spotted a large transport ship berthed in the military zone. He fired off one torpedo but its rudder locked hard over and it ran amok. It started going round the harbour in huge circles – even threatening E11 herself. Nasmith rapidly ordered the second bow torpedo to be fired and it went straight for the target and the crew listened as it struck the ship. Nasmith said he was unable to 'observe the effect' through his periscope because E11 was being swept around by cross-tides: he was also concerned about colliding with the rogue torpedo.

'Full ahead both! Seventy-five feet,' he shouted. 'Flood the auxiliary. Take her down, fast!' There followed several minutes of nerve-racking drama for Nasmith and his crew. The infamous Bosporus currents and counter-currents took command and E11 swung this way and that and finally thumped against the bottom of the strait. She stayed motionless for a moment then bumped forward and up a bank and then after another bounce came to rest only a few feet below the surface. One more bounce and she'd be revealed to the shore batteries and exposed to devastating gunfire. There was nothing Nasmith could do. Then, miraculously, the submarine started to swing. Other currents were nudging her from the incline on which she was stuck. Nasmith noticed the compass needle moving. The submarine was working her way free. The moment she was floating Nasmith prayed he was heading the right way and ordered a dive. When he next raised the periscope the dome of the great mosque of Santa Sophia was way off to the north-west. E11 was free. She headed for the centre of the Marmara where all seemed quiet. The crew spent the next day enjoying the fresh air and having another swim.

Nasmith had now destroyed four Turkish vessels. He had only five torpedoes left. He decided to make a very risky move in order to preserve what firepower he had. He modified his remaining torpedoes to stay afloat if they didn't explode. 'We can't afford to lose a single one of them,' he said. 'If we miss, we'll pick them up again.' His work soon proved worthwhile. On 28 May smoke to the north-east signalled the presence of a large supply convoy heading towards them. Nasmith lined up E11 so that he had the largest of the ships in his sights. At 7.30 a.m. he fired a torpedo from his port beam torpedo tube. It struck the ship towards the stern, and, as E11's navigator wrote in his log: 'The stricken vessel seemed to be completely lifted out of the water aft, and heeling over to starboard, sank in less than a minute after being hit.' Four and a half hours later Nasmith and his crew fired one of his remaining four torpedoes at a steamer heading west out of Constantinople: she appeared to be loaded with Turkish troops heading for Gallipoli. Nasmith watched as his torpedo seemed to strike the ship but then failed to explode. His frustration turned to relief minutes later when a closer look at the decks through his periscope told him that the ship was crowded with civilians not troops.

But where was that torpedo and why had it failed to explode? Once the steamer had gone and the horizon was clear the crew searched the area for the stray torpedo. It didn't take long to find it floating on the surface. There was no small boat aboard the E11, so Nasmith himself stripped down and dived into the sea, telling the crew not to allow the submarine to get close to the seventeen-foot-long torpedo until he had neutralised it. He asked for a spanner and swam to it. He immediately saw the top of its head was crushed in. It had struck the ship but somehow failed to explode. Very carefully he unscrewed the firing pistol, which contained the pin that would strike the detonator, and set it to 'Safe'. He then swam back to E11, manoeuvred her alongside the torpedo and had it hoisted on board. The fore hatch was opened and the warhead detached to allow the body of the torpedo to be lowered into the hull. They were nervous minutes. If an enemy ship appeared on the horizon, the whole crew knew the submarine couldn't dive with the fore hatch open. But after this successful recovery Nasmith once again had four torpedoes ready to fire.

Three days later, after picking up some fresh food supplies from a Turkish yacht they had stopped and searched, Nasmith and his crew torpedoed a large ship embarking troops in a harbour on the south side of the Marmara. She exploded instantaneously, obviously full of ammunition, and two days later another store ship was sent to the bottom. Two torpedoes left. Within three hours one of them was fired off at another store ship, but it missed and soon appeared floating on the surface. Nasmith's number two, D'Oyly Hughes, persuaded his skipper to allow him to jump overboard and defuse it. This time they managed to stuff the torpedo into its tube from the outside, avoiding the risk of being caught with the fore hatch open.

Nasmith might have stayed on patrol longer but he discovered there was a crack in the submarine's starboard drive shaft. He set a course back to the Dardanelles, calculating that he would probably find a use for his two remaining torpedoes on the way home. At noon on 7 June they caught a large troopship loitering halfway down the strait and torpedoed and sank her. They were almost home after a patrol that was already fast becoming a triumph. But their most unsettling trial still lay ahead.

Just as their long underwater escape from the Dardanelles reached the most dangerous spot where Turkish defences were strung across the waterway at its narrowest point, there was a sharp clunk. They couldn't be aground. It was too deep. Nasmith ordered E11 up to periscope height then gingerly edged its top above the surface. What he saw filled him with horror. A huge horned mine had come free of its moorings and was being dragged along by the submarine, its hawser caught by one of the hydroplanes, the adjustable fins that controlled E11's angle in the water. If one of the horns caught the surface of the hull, they'd all be blown to bits. With his eyes still at the periscope, Nasmith decided not to tell anyone what was happening. For what seemed an age he guided the submarine through the rest of the minefield and out of the strait to a place at the southern end of the Dardanelles where they could surface within reach of British ships. The mine was still floating above the bow, its horns miraculously staying clear of contact with E11's hull. Nasmith ordered his men to blow air into the submarine's after tanks and keep the bow tanks full of water. Then with the bow still below the surface and the mine floating above, he shouted, 'Full astern both!' The submarine heaved backwards and the mine's mooring line slipped off. 'Leaving the bow submerged,' Nasmith wrote in his log, 'and bringing the stern to the surface, the rush of water from the screws, together with the sternway gathered allowed the mine to fall clear ahead of the vessel.' He watched the lethal ball of metal float away and then disappear beneath the surface way behind them.

Nasmith's patrol had been behind enemy lines for three weeks and destroyed nine Turkish vessels, one of them in Turkey's main harbour. He had one torpedo left. He had been keeping it just in case he spotted the great Turkish battleship *Barbarossa* on his way out of the straits. But Nasmith had to wait for his chance to get at her. The Royal Navy was so impressed by E11's actions that it sent Nasmith and his submarine back into the Sea of Marmara twice more. *Barbarossa* was among the first targets destroyed on the second patrol. Yet victory over the Turks was to remain out of reach.

In spite of the robust support of submarines like Nasmith's, which survived unlike many others, the allied army's invasion of Gallipoli was a tragic and costly catastrophe. The assault, which began in April 1915, was abandoned in January 1916 with a loss of nearly 190,000 allied troops during the campaign.

Nasmith won the Victoria Cross for 'most conspicuous bravery'. His two subordinate officers, Guy D'Oyly Hughes and Martin Brown, the navigator, were awarded the Distinguished Service Cross and each of the rest of the crew a DSM, the Distinguished Service Medal. Nasmith rose to be an admiral by the Second World War and died in 1965 at the age of eighty-two.

Martin Nasmith and his crew had earned their medals. By their own count, their three patrols into the Sea of Marmara had cost the Turks twenty-seven steamers and fifty-seven smaller vessels

23

Charles Lightoller

From the Titanic *to Dunkirk, 1912–40*

IT WAS 1 June 1940. Under a clear blue sky Commander Charles Herbert Lightoller gave the command for his son, Roger, to steer their motor yacht *Sundowner* out of Ramsgate harbour. She was their pride and joy – an old wooden steam packet boat that Lightoller had powered with a seventy-two-horsepower diesel engine. He'd spruced up the vessel with a couple of masts, a wheelhouse and roomy cabins below. Through the 1930s he and his wife, Sylvia, took family and friends on holidays the length and breadth of the North Sea. He and *Sundowner* were inseparable: 'If anything ever happens to me,' he told Sylvia, 'just put my body on board, tow her out and set fire to her.'

Then came the Second World War and the critical summer of 1940. When the news reached London that the British Expeditionary Force of some 400,000 men was trapped by Hitler's army in and around the Channel port of Dunkirk, the word went out that every suitable vessel big or small was urgently needed in a desperate attempt to save them. The Admiralty notified Lightoller that a naval crew would take over *Sundowner* to go to the army's rescue. No way, retorted Lightoller, if anyone was to take her across the Channel it would be himself and his son. 'No problem,' said the navy, delighted that they'd have one less boat to crew. 'Off you go!'

As Lightoller, Roger and a friend, eighteen-year-old Sea Scout Gerald Ashcroft, steamed out of the harbour and headed south-east, a huge black smoke cloud rose on the horizon ahead of them accompanied by the unmistakable sound of war.

Sundowner's skipper was no pleasure-boat enthusiast eager to prove himself. He was a retired naval commander who'd had a distinguished career. Lightoller was sixty-six years old and this was to be the last and most dangerous challenge in a life of unparalleled adventure at sea. He had survived several shipwrecks including, most remarkable of all, the *Titanic* in 1912, and fought his way with the Royal Navy through the First World War. One destroyer he commanded rammed and sank a German U-boat in the North Sea. With her bow shattered, his ship looked like sinking too but Lightoller skilfully manoeuvred her back to her base in the Humber.

By the time he set course for Dunkirk in 1940 Commander 'Lights' Lightoller was already one of the most famous seamen afloat. Many believed he had a charmed life. He first went to sea at the age of thirteen in the days of the great sailing ships. One of his luckiest escapes was when his ship, the *Holt Hill*, had the misfortune to collide with one of the remotest islands in the world, tiny St Paul in the southern Indian Ocean in 1889. By an extraordinary fluke a passing British barque, whose captain was curious to inspect the island, came near enough to spot the desperate signals of the shipwrecked crew and rescued them. This was only one of Lightoller's early scrapes. He survived another of his ships being raked by fire at sea, another losing her masts and a frantic but unsuccessful scramble to save four crewmates who were drowned when his gig was overturned by a huge wave. In his memoirs he recalled with typical nonchalance: 'The sea tried to drown me several times, yet I beat her . . . we still remain the best of friends.'

But of all his adventures at sea, none before Dunkirk compared to the night the *Titanic* hit an iceberg. Lightoller became a maritime celebrity for his heroic actions – herding women and children into lifeboats on the sinking ship. He joined the prestigious White Star Line and its grand ocean liners in 1900. Twelve years of experience and promotion led him to be appointed second officer on the brand-new and massive *Titanic*. 'It took me fourteen days before I could with confidence find my way from one part of the ship to another by the shortest route,' he recalled. He found himself close to the captain, E.J. Smith, 'E.J.' as he was affectionately known. Lightoller described him as 'tall, full whiskered and broad'. He expected Smith to have a 'voice like a foghorn' but found his captain was actually softly spoken and had a constant smile.

On the fateful night of 14 April 1912 at 10 p.m. Lightoller, on watch, handed over to the first officer William Murdoch. 'We were then making an easy 22 knots. It was pitch dark, and dead cold. Not a cloud in the sky, and the sea like glass.' It seemed plain sailing but an urgent radio message from another ship in the area that she had seen a great number of large icebergs was not passed on to the bridge. Lightoller said it was this failure to sound the iceberg alert that was 'the main contributory cause to the loss of that magnificent ship'.

It was unusually cold and Lightoller, now off duty, went back to his cabin and wrapped himself tightly in his blanket. He was just falling asleep when he felt a sudden jar and a powerful vibration that ran right through the ship. Minutes later the fourth officer knocked at his door. 'We've struck an iceberg,' he said, 'and the water is up to F Deck in the Mail Room.' Lightoller leapt out of bed. F Deck under water was serious, although he didn't at this stage believe it would prove fatal. The engines stopped and the surplus steam gave out a huge roar which only added to the anxiety of the passengers who began flocking up to the boat deck. Lightoller, smiling encouragement to them, began ordering the crew to strip the covers off the lifeboats. He was convinced the ship would not sink. He believed that a couple of *Titanic*'s forward compartments had been holed: in fact the sea was pouring into no fewer than six of them.

Lightoller found Captain Smith and asked him whether he should load up the lifeboats. When the captain said yes, Lightoller asked him: 'Women and children away, sir?' Captain Smith nodded without speaking. Astonishingly *Titanic* had lifeboats for only 1,100 of the

2,200 people she was carrying. And there had been no lifeboat drill since the liner left Southampton four days earlier. In his rigorous supervision of the loading of the boats, Lightoller refused to allow men to board them. At one point he waved his revolver and threatened to shoot a group of men who'd jumped into one boat. 'They hopped out mighty quickly . . . I had the satisfaction of seeing them tumbling head over heels on to the deck . . . the revolver wasn't even loaded.' He occasionally lowered boats away before they were fully loaded with women and children even when men were available to clamber in. He was later criticised for contributing to the loss of life by sending off some boats partly empty. Lightoller feared that if he filled the boats when they were still slung high up in their davits, the people in them might be at risk of falling out; he planned to complete the loading once the boats had been lowered to a safer level. But the fast sinking ship made that impossible. The irony was that Lightoller and other officers were never told that the lifeboats and their davits were designed deliberately to allow a full complement to be loaded at the top level. *Titanic*'s safety procedures and her crew's knowledge of them were desperately inadequate.

One thing that kept Lightoller's hopes up was the sight of a ship's lights not far off. These were the lights of the *Californian*, but her radio operator was not at his receiver throughout these vital few hours. Some of *Californian*'s officers saw *Titanic* firing white flares, but her captain said they were of no importance and steamed on. His and his radio officer's negligence left 1,500 people to drown including Captain Smith who went down with his ship.

As *Titanic* foundered, Lightoller did what he could to fill the lifeboats. He found one couple sitting together on the deck. 'Let me put you in one of the boats,' he said to the woman.

'Not on your life,' she replied. 'We started together and if need be we'll finish together.'

As one of the very last boats was lowered, two men jumped into it. 'I don't blame them,' wrote Lightoller later, 'the boat wasn't full, for the simple reason we couldn't find sufficient women. Good luck to them.'

There was now nothing Lightoller felt he could do to help anyone else. He watched the stern rising faster and faster out of the water and the doomed crowds of people still on the ship trying to claw their way up the deck. 'I knew all too well the utter futility . . . of struggling up towards the stern . . . There was only one thing to do, so, turning to the fore part of the bridge, I took a header. Striking the water was like a thousand knives being driven into one's body.'

The water was four degrees below freezing. Lightoller struck out, ditching the Webley revolver that was weighing him down, and swam frantically to get away from the downward surge of the sinking ship. He was surrounded by struggling people 'some swimming, others mostly men, thank God, definitely drowning – an utter nightmare of sight and sound'. He managed to make it to an upturned boat and somehow hauled himself up so that he was standing on its keel in full view of *Titanic* as she finally slid beneath the sea. 'With impressive majesty and ever increasing momentum she silently took her last tragic dive.'

'Lights' was the most senior crew member to survive, and he was soon thrust into the forefront of the round of cross-examinations that followed. He found the experience uncomfortable and the limelight distasteful and was grateful to be able to join the Royal Navy. It gave him a number of challenging commands during the First World War. His most notable exploit was when his destroyer, HMS *Garry*, rammed and cut in half a U-boat that had surfaced just ahead of her.

Lightoller briefly returned to the White Star Line but it soon became clear to him that the shadow of the *Titanic* disaster would always deny him senior command. To the management Lightoller, like the handful of other surviving crewmen, was bad for business. He resigned and spent the 1920s and 1930s quietly pursuing a range of businesses and a happy home life. But he never lost his love of the sea and boats. And *Sundowner*, the elegant, sharp-nosed motor launch he bought between the wars, became his prize possession. In May 1940 *Sundowner* and her skipper, by now in his sixties, were a natural choice for the admirals seeking 'little ships' in the first dire emergency of the Second World War.

As *Sundowner* approached Dunkirk, Lightoller's son Roger was at the wheel. Lightoller watched the sea ahead with eagle eyes. 'Hard a starboard!' he suddenly shouted at his son. *Sundowner*, unlike a giant liner, had a useful habit of responding instantly to a shift of her rudder. Roger didn't waste a second. The boat swung sharply to the right and her crew watched with relief as the horns on the top of a live German mine floated gently by a few feet to their left.

As their boat approached the shore, they were plunged into the full horror of battle and faced the threat of destruction from Nazi gunfire and dive-bombers. Shells exploded all around them, the noise was deafening, the sky darkened by smoke. Lightoller spotted another motor cruiser on fire a little way off and drew alongside to rescue her two crew members and the three sailors she'd picked up at Dunkirk. 'Move off fast,' advised her crewmen once they were aboard: there were 200 gallons of fuel on the boat. Moments after *Sundowner* pulled away, the abandoned vessel blew up.

Lightoller's other shipmate, the young Sea Scout Gerald Ashcroft, had helped Lightoller strip *Sundowner* for action back in the Thames. He was delighted to be asked to go on the rescue mission, thrilled at the prospect of seeing the war first-hand. He was fortunate to be with a very experienced skipper. 'We attracted the attention of a Stuka dive bomber,' he later reported. 'Commander Lightoller stood up in the bow and I stood alongside the wheelhouse. Lightoller kept his eye on the Stuka till the last second – then he sang out to me "Hard a port!" and I sang out to Roger and we turned very sharply. The bomb landed on our starboard side.'

When Lightoller spotted a fighter approaching from astern, he knew that just a second or two before it fired its guns, it lifted its nose. 'Just as he thought the plane was about to lift,' said Ashcroft, 'he sang out to go hard a port, which we did – and the bullets came flying down the side.'

Sundowner headed in towards the main harbour jetty at Dunkirk. Some 200,000 men had already been rescued, but there were 200,000 more to go. In the harbour there was debris everywhere. Bodies and floating wreckage threatened to get caught up in *Sundowner*'s propeller. Young Gerald Ashcroft was given the job of fending them off with the boathook.

Because the tide was low, Lightoller told Roger to berth alongside the British destroyer *Worcester*.

'How many can you take aboard?' shouted the warship's commander.

'About a hundred,' replied Lightoller. In the past more than twenty passengers had been a crowd but *Sundowner* had been stripped of all her furniture and bunks and there was room for more. In the end, as *Worcester*'s guns provided cover from the constant air attacks, no fewer than 129 men came aboard, some of them wounded. Two even squeezed themselves into *Sundowner*'s bath.

As she pulled away from the destroyer and headed back out to sea, Lightoller's vessel was attacked again and again. One of the most frightening moments was when a Luftwaffe bomber headed straight at her. Because of Lightoller's fast reactions it twice failed to hit the small boat. A British Spitfire providentially roared in and sent the German plane down in flames only fifty yards behind *Sundowner*.

At 10 p.m., twelve hours after she had left Ramsgate harbour, *Sundowner* and her precious cargo were back. The rescued men were so excited at the sight of the British quayside that there was a rush to get on deck and the boat nearly turned over. Once they were disembarked there appeared to be one man missing. Lightoller went below to search and 'there, sitting on the loo, was one nearly unconscious Tommy,' recalled Ashcroft.

'Come on, lad,' shouted the skipper, 'time to go home.'

The Second World War was also to bring tragedy for Lightoller and his wife Sylvia. Two of their three sons died in the conflict. Roger lost his life just one week before VE Day. Charles Herbert Lightoller himself died in 1952. Sylvia kept *Sundowner* in commission and in 1990, fifty years after Dunkirk, the motor yacht led the procession of surviving Little Ships across the Channel to commemorate their famous voyage. *Sundowner*'s skipper was Captain Tim Lightoller, a Royal Navy submariner, and grandson of the old sea dog. One of his passengers was the former Sea Scout Gerald Ashcroft. 'We were given a huge welcome in Dunkirk,' Lightoller told us. As a boy of ten he'd accompanied his grandparents on several voyages in the late 1940s. 'My grandfather was a man of few words. He was not the man to boast about the past, but he taught me everything I know about seamanship.' *Sundowner* also proudly took part in the Queen Elizabeth's Diamond Jubilee parade of Little Ships in the Thames in 2012.

WAR IN THE AIR

THE COURAGE OF seamen in wartime is exemplary. But it takes a very special kind of courage to be a wartime fighter pilot. The history of air combat is relatively recent. The aeroplane was only invented a decade before the First World War. It was to transform the way battles were fought. The first pilots were observers, sent up to scan the battlefield from above and report back on troop movements. But it wasn't long before they were fighting against enemy pilots and attacking targets on the ground. They became national heroes for their flying agility and daring and there was fierce competition between them to chalk up high scores of enemy 'kills'.

All but a very few of the young men who volunteered for the cockpit would be dead within days. Inexperience killed. One moment's lapse of concentration and you were easy prey. In 1915, during a period when the Germans had air supremacy, the average life expectancy for allied pilots was just seventeen and a half hours of flying time. The average lifespan for a Spitfire fighter pilot during the Battle of Britain in 1940 was only four weeks. And yet the recruits kept on coming, lured – like Yeats's Irish airman – by the fascination of a machine that offered escape from the tyranny of the trenches on the ground.

> Nor law, nor duty bade me fight,
> Nor public man, nor cheering crowds,
> A lonely impulse of delight
> Drove to this tumult in the clouds

If you were lucky, very lucky, you'd survive your first dogfight, and earn yourself fame as a fighter 'ace'.

Control of the air is today a vital ingredient of any country's military strategy. No other weapons can destroy an enemy with the reach and immediacy of attack from the air. And the human pilot can now be replaced by drones and missiles travelling hundreds of miles in minutes and guided to their targets by operators snugly secure hundreds or thousands of miles behind the lines. Compared to the 'pilots' of this modern wizardry the two men whose stories we've chosen to tell had to live under constant threat of sudden death. Both led the way in aerial combat in its heyday. Canadian Billy Bishop was a hero in the First World War. His story is a fascinating study of technical skill and courage bordering on foolhardiness. The same is true of our Second World War choice – the American pilot Butch O'Hare. Both strike us as men who made an indelible mark on military history.

24

Billy Bishop

Canadian Fighter Pilot, 1915–18

ONE DAMP DAY in 1915 two young First World War army recruits waded through the mud of their cavalry camp. The war had been raging for a year and no end was in sight. The men were cursing the persistent drizzle that had turned the ground into a claggy slime that sucked at their boots. They struggled to take each step. One of them lost his balance and fell face down in the muck. The other was about to reach down to help, when he spotted a small aircraft flying overhead and landing carefully in the next field. He later recalled watching 'the trim little aeroplane' take off again and disappear 'into the clean grey mists'. Turning to his friend he said, 'George, it's clean up there. I'll bet you don't get any mud or horseshit on you up there. If you died at least it would be a clean death.' As they slogged their way back to the camp, the twenty-one-year-old Canadian volunteer, whose name was Billy Bishop, told himself: 'There is only one place to be on such a day – up above the clouds and in the summer sunshine. I am going into battle that way. I am going to meet the enemy in the air.'

Several weeks later Bishop was interviewed about his request for a transfer to the Royal Flying Corps – the RFC – then still in its infancy. Orville and Wilbur Wright had made the world's first flight only twelve years earlier. Some military chiefs scoffed at this new invention: what could aircraft do in war? The general who ran the British army back in 1910 called the whole notion of military aviation a 'useless and expensive fad'. But within weeks of the beginning of the war in 1914 the new flying machines were proving invaluable, monitoring battles from above. Their pilots and observers could see far into enemy territory and report back on hostile movements. They could spot targets for their own artillery batteries and report whether the shells were finding their mark. And, it rapidly became clear, aircraft were a new form of weapon. They could attack the enemy on the ground with guns: even hand grenades thrown out of cockpits did telling damage. It didn't take long before those grenades were replaced by bombs. And as the RFC developed, the fighter aircraft (called the 'scout') became a vital weapon in seeking out and destroying enemy planes. The RFC quickly proved its worth and became the Royal Air Force in 1918.

Billy Bishop attended his interview with the RFC a year after the war began. The perky lad, whose only flight had been a near fatal one in a cardboard box from the third storey of the family house in Owen Sound, 100 miles north of Toronto, was questioned by a moustachioed Old Etonian. Can you ride a horse? Yes, replied Bishop, wondering what this had to do with flying aeroplanes. After establishing that Bishop could ski and skate and drive a car, all of which apparently suggested he had the sense of balance to fly the RFC's primitive aircraft, the bigwig told him he could join but only as an observer. Bishop leapt at the opportunity. He was soon learning to fly, although it took him some time to become a skilled pilot. He damaged a number of aircraft in training and didn't get his own pilot's seat until the end of 1916. But he finally delighted his instructors with his agile handling of the rickety biplanes – the ones with two wings, stacked one above the other. He had two other qualities that signalled a promising career as a fighter pilot: he was utterly fearless and he had proved himself a marksman hunting pigeons with a .22 rifle his father gave him as a boy.

In March 1917, Billy Bishop flew his Nieuport biplane across to France to begin the spectacular series of successful combats in the air that would establish him as the greatest of all allied First World War aces. His list of successes was so spectacular that some began to doubt his claims. But no one disputed that he was one of the very best. The young Canadian arrived at the French base at Filescamp near Arras when the RFC was at a low ebb – at the darkest moment in the whole war. The Germans appeared to be winning. The British and French had suffered more than half a million casualties on the Somme and at Verdun. Russia, racked by revolution, was collapsing on Germany's eastern front. The life expectancy of the RFC's pilots was on average three weeks. The Germans were winning more and more dogfights with their lethal new Albatross fighters – fifteen miles an hour faster than the RFC's Nieuports. Only a few miles away across the front lines the most fearsome Albatross pilot of them all, Baron Manfred von Richthofen – known as the Red Baron because his aircraft was painted fiery red – had already dispatched thirty British planes. Bishop couldn't wait to confront him and the German air force – the Luftwaffe – which was casting a dangerous shadow over the battlefields in northern France. With the war in the trenches on the ground at a virtual stalemate, supremacy in the air was more important than ever.

Major Jack Scott arrived at Filescamp the day after Bishop to take over command of 60 Squadron. Scott had been crippled in a crash in 1914 but although this left him hobbling across the tarmac to his aircraft, once in the cockpit he was one of the best squadron leaders of the First World War. Under Scott's leadership Billy Bishop thrived. The air was a new world for warriors used to fighting on land or water – a world of aerobatic mortal combat that demanded split-second reactions and a murderous will to destroy your opponent. Flying the primitive contraptions built of wood and canvas was also a dizzying physical challenge: the aircraft were slow, the quickest could move no faster than 130 miles an hour, and the pilot, sitting in an open cockpit, was exposed to lacerating wind and unpredictable weather conditions. The only sure way to survive an attack was to be ready to put the aircraft into a dive or spin that risked loss of control. A further hazard was that the pilot sat within inches of a fuel tank that one bullet could explode. It's hardly surprising that the death toll among pilots in the RFC was proportionally far higher than for infantry officers fighting on the ground.

Billy Bishop seemed utterly immune to all the risks and dangers of flying. And he never doubted that his job was to kill, to see 'the Hun' crashing down in flames in front of him. 'It was such a wonderful way to take life,' he wrote, 'that, looking back at it, I feel that nothing the future can ever hold for me can excel those wonderful days . . . face to face with death every day.'

Bishop's first real fight in the air was on 25 March. He climbed into his single-engined Nieuport biplane and was soon flying in diamond formation with Jack Scott in the lead. There was low cloud to start with which protected them from the anti-aircraft barrage aimed at them as they crossed into German-held territory. They found a spot where the Germans had pulled back a little. 'After we had flown well into what had been so recently Hunland, the weather suddenly cleared.' And out of the corner of his eye, Bishop spotted three German aircraft approaching from behind. That spelled lethal danger, and what was more they were Albatrosses, the best planes the Germans had, far more streamlined than the British Nieuports, shaped like torpedoes and with large black Maltese crosses painted on their sides. Bishop could hardly believe that Scott allowed the Germans to come as close as 400 yards to his rear. But Scott was luring them into what he hoped would be a trap. Suddenly he swung his aircraft round and the others followed, Bishop a little slower than the rest. The trick was to bank so fast that the enemy who had been on your tail was now in front of you – a prey to your machine guns. But when Bishop turned he saw one of the Germans diving straight at Scott's aircraft. Quick as a flash he aimed directly at the Albatross and found he could shoot right at the pilot's back. He fired forty or fifty rounds from fifty yards behind. Many of the tracer bullets hit the German's fuselage and at least one seemed to hit the pilot. The plane plunged into a spinning nosedive. It could have been a trick. Bishop dived after it just in case. Seconds later he watched it crash.

It was his first 'kill'. But he had no time to congratulate himself. 'To my dismay, I discovered that during our long dive my engine had filled up with lubricating oil and had stopped dead still. I tried every little trick I knew to coax a fresh start but it was no use.' Frantically, he scanned the ground below him to try to discern if it was enemy territory. The rattle of a machine gun soon told him that it was. 'I continued to glide, listlessly, toward the ground, not caring much now what the machine gun might do. My plight couldn't be much worse.' Ironically he was saved by the fact that those crude early aircraft could do what today's warplanes cannot: they could glide. He drifted on for a while and managed to aim his Nieuport at a clear spot in the rough terrain. He landed, leapt out of the plane and dived for cover. Moments later he peeped over the edge of the bank in front of him and saw four men crawling towards him. It was only when they were almost upon him that he recognised their khaki uniforms and realised they were British. 'The Tommies had seen me and had bravely crawled out to help me.' He was in the middle of no-man's-land, just 150 yards from German trenches. He was escorted back to the British lines and a few days later his plane was dismantled and driven back in a truck to the airbase. 'I was very proud . . . and excited over the whole episode.' He wrote a letter home describing it all as 'great sport', adding, 'I never enjoyed myself so much in my whole life.'

It may have been great sport for him but in the first five days of April 1917 the Royal Flying Corps lost no fewer than 75 aircraft and 103 airmen. Like all the other allied forces it was stretched to breaking point but Billy Bishop saw this as an opportunity. He flew almost every day. 'Some days I could have been accused of violating all the rules of a flying man's union (if we had had one). I would fly as much as seven and a half hours between sunrise and sunset. Far from affecting my nerves, the more I flew the more I wanted to fly.' Some of his most notable victories were ones he fought alone. He mastered the technique of creeping up on his victims from behind – either from way above them or just below them, where they were least likely to spot his approach.

As his luck held and he survived combat after combat, Bishop chalked up scores of victories over the Luftwaffe that began to threaten the records of other First World War aces who'd been at it longer than he had. Bishop made no secret of his ambition to pile up a record-breaking total, and to win himself a chestful of medals. He was also developing an unattractive streak that made him sound like a psychopath. 'I must say that to see an enemy going down in flames is a source of great satisfaction,' he said. And in a letter to his girlfriend, Margaret, whom he was later to marry, he wrote: 'You have no idea how bloodthirsty I've become and how much pleasure I get in killing Huns.' The strain on a man dodging death every time he took to the cockpit, sometimes three or four times a day, must have been extraordinary. He let his hair down liberally when he was on the ground. He was notorious in the officers' mess for enjoying wild parties and – in spite of his lasting devotion to Margaret – for the occasional amorous adventure.

As the Canadian pilot ruthlessly piled up a string of successes, the Luftwaffe – in particular the Red Baron – made Bishop's destruction a priority. On 30 April Bishop was caught up in a tussle with a flight of four red Albatrosses. He was convinced they were Richthofen and three of his best men. 'I opened fire on the baron, and in another half moment found myself in the midst of what seemed a stampede of bloodthirsty animals. It was a lightning fight, and I have never seen anything like it.' Bishop escaped and after he landed he found seven bullets had passed 'within an inch of me in one place'.

Bishop's most controversial raid was the one that won him the Victoria Cross. At a party in the officers' mess on 1 June 1917 he revealed that he was bent on going in at dawn to strafe a German airfield and pick off the pilots one by one as they tried to get airborne. It was an outrageously risky plan. German aerodromes were protected by clusters of anti-aircraft guns and one or two of the Germans would surely manage to take off and attack him. Bishop's comrades in the mess told him he was mad and none agreed to join him. One of them told him it was 'sheer suicide'. Bishop insisted they were wrong: 'Nothing like it had been done before. I knew that I would strike the Huns by surprise . . . the risk was not nearly as great as it seemed.'

Bishop told his batman to wake him at 3 a.m. The weather was poor but he decided to go ahead. His Nieuport was wheeled out, his mechanic swung the propeller and the engine burst into life. Bishop checked the gun was loaded. It wasn't on a wing: a remarkable wartime invention allowed the gun to fire between the blades of the propeller. Considering how fast the blades moved, it was an astonishing miracle of synchronisation. Once in the air Bishop crossed the German front line and, disappointed to find no aircraft on the ground at the airfield he had targeted, he flew on to find another one further behind German lines. He saw seven planes on the ground with pilots and mechanics standing around. He flew over them at no more than fifty feet, spraying machine-gun bullets all over the apron. 'Then clearing off to one side, I watched the fun . . . they tore around in every direction like people going mad or rabbits scurrying around.' He spotted one aircraft race for the runway and start to take off. He managed to get close on its tail and open fire. 'There was no chance of missing and I was as cool as could be.' The German plane slipped out of control and crashed. Bishop chased another plane along the runway. It just got off the ground but the pilot dipped to avoid Bishop and hit the trees. Another two planes then took off and Bishop fired at and destroyed one and frightened off the other. He was back at base by 6 a.m. 'I have always kept the telegrams of congratulations which I received that day,' he wrote.

Some of Bishop's colleagues had their doubts about his claims, but the government was looking for war heroes. Two months after his attack on the airfield, King George V decorated Billy Bishop with the Victoria Cross. He also received the Military Cross for his destruction of a German balloon and the Distinguished Service Order for another successful combat. 'You have been making a nuisance of yourself out there, haven't you?' said the king, adding that he'd never before presented 'all those three honours for gallantry to one person'.

For the best part of a year after the investiture Bishop was kept away from the front. The government wanted to preserve its heroes and use them to promote their stories and win recruits. But Bishop was desperate to get back into the air. And in April 1918 he heard that the Red Baron – von Richthofen – had been shot down and killed – and he had missed it! Within weeks he was back in the cockpit of the fastest new British warplane – the SE5A. He was now married to his Canadian sweetheart Margaret. She knew she was unable to persuade him to stop risking his life, so she asked his fellow pilots to 'stick to the Major [his new rank] and not let a Hun get on his tail'. In the course of May and June Billy Bishop's victories reached almost as high a level as the Red Baron had notched up in the whole war. But on 15 June came the most unwelcome news of his career. The Canadian government, anxious that their top ace should survive the war, recalled him to Canada, promising him the job of forming a new Canadian air force. 'I have never been so furious in all my life,' wrote Bishop. Given permission to fly until the final call came, he spent the next two days destroying five German aircraft.

Late on the second day the final summons came to leave the war: 'You're to fly back to Britain by noon tomorrow.' That night was spent farewell drinking until 3 a.m. But as the party ended he was heard to moan, 'Oh, for one more fight in the air!' An hour or two later, seriously hung-over, twenty-four-year-old Billy Bishop risked one last flight. Within minutes he was over German-held territory with three Luftwaffe fighters in front of him and two on his tail. One burst from his guns sent the nearest of the three ahead of him down in a nosedive. The other two swerved to avoid him, collided with each other and plunged into the forest below. Instantly he jerked the joystick backwards and threw his plane into a tight loop to latch on to the tails of the two Germans who'd been behind him. One of the aircraft crashed in flames, the other managed to escape. But Bishop wasn't finished: his last flight was to claim one more victim. He spotted a two-seater reconnaissance aircraft and swept under it, firing twenty rounds into its underbelly. The wounded pilot crashed into a hillside. It was Bishop's seventy-fifth claimed 'kill' – just five short of his arch-rival, the record-breaking Red Baron.

Billy Bishop lived long after the end of the war. As an air marshal he went on to lead the Royal Canadian Air Force that he had set up in the Second World War and when he died in 1956 at the age of sixty-two, he gave his name to Toronto City Airport.

25

Butch O'Hare

Second World War Ace, 1941–3

TAKE A WALK through the main terminal at Chicago's busy airport and you'll come across an unexpected sight – a Grumman F4F Wildcat fighter plane. It's an exact replica of the one used by Edward Henry 'Butch' O'Hare, a Second World War naval air ace after whom the airport is named. The world of the aerial dogfight is full of heroes, but we've chosen to tell Butch O'Hare's story because he's in a class of his own. He single-handedly saved a US aircraft carrier from destruction. That won him the first naval Medal of Honour for 'one of the most daring, if not the most daring single action in the history of combat aviation'. He also has another claim to fame: a connection to one of America's most notorious criminals, Al Capone.

Butch O'Hare started life as he ended it – as a risk-taker. He took to parachuting early – jumping off his family's garage roof with an umbrella. True, he jumped into a snowdrift but the landing was so hard he didn't try it again. He loved racing around on bicycles, the faster the better. He was a natural sportsman and played American football which he later claimed made him a good team player – quick to react under pressure. He was also president of his high school's rifle club which helped prepare him for his extraordinary skill as a fighter pilot.

Butch was still at school when his father, Edgar 'E.J.' O'Hare, got mixed up with the mob. A lawyer turned businessman, O'Hare senior bought a racetrack in Chicago and became an associate of the legendary gangster Al Capone. Then, fortunately for the family's reputation, he turned whistle-blower. In a bold switch of loyalty he offered to work undercover for the US tax authorities. Butch's father led them to Capone's accountant which helped supply the evidence they needed to charge him with tax evasion. He also proved instrumental in ensuring a fair trial. In October 1931 O'Hare Senior warned officials that Capone's henchmen were intimidating the jurors and on the first day of the trial the presiding judge replaced the nobbled jury with another one. Capone was sentenced to eleven years in jail. But E.J. O'Hare was to pay dearly for his double-cross. In November 1939, while driving through Chicago in broad daylight, he was shot in the head and died instantly. His assassin escaped but the shooting bore all the hallmarks of a Capone gangland slaying.

There are many theories as to why Edgar O'Hare fingered Capone. The most convincing one is he did it in exchange for government help to get his son Butch into a good military school. One of the US officials involved in the case later wrote that, after Capone was convicted, O'Hare's 'big dream was realised. His boy Butch did receive an appointment to Annapolis Naval Academy from which he graduated with high honours.'

At Annapolis Butch O'Hare was a popular member of the Class of 1937. He was now a handsome twenty-three-year-old, five foot ten and a half, with neatly combed black hair. A roommate described him as a 'hulking figure' who was 'self-assured and exuded confidence'. He made it clear that his aim was to become a naval pilot.

O'Hare's flight training started officially in June 1939 in Pensacola, a US naval base in Florida known as the 'Cradle of Naval Aviation'. He was the first in his class to make a solo flight. An instructor noted: 'Student handles plane more like an experienced pilot than a student. All landings and take-offs very consistent and good.'

After a year of training, his flying skills marks were 'above average'. His aerial gunnery won him an 'excellent'. His instructors reported that Butch O'Hare already showed qualities that would serve him well in combat. One wrote 'I consider this officer's judgement above average. He thinks clearly, is thorough in his work and is dependable. He appears to be unexcitable and should make a very satisfactory pilot.' The highest praise came from a fellow trainee who wrote: 'We all envied him, in a sense, because an airplane was made for him or he was made for an airplane. From the moment he stepped into an airplane, nothing went wrong. He had a touch that was unbelievable. He was just a completely natural pilot, just tremendous.'

On 1 July 1940, proudly sporting his new gold wings on a blue uniform, O'Hare reported for duty to Fighter Squadron Three based on the aircraft carrier USS *Saratoga*. She was one of the first of the great vessels to project military power at sea. O'Hare was joining a force that would soon be locked in a titanic struggle with Japan for mastery of the Pacific Ocean. And as a pilot of a naval warplane he would be in the front line of the aircraft carrier's unique power to strike an enemy way beyond the guns of any other kind of warship. He would fly the newest version of the Grumman Wildcat fighter. Highly manoeuvrable, it was for its time a fine naval warplane with a speed of 330 mph – almost but not quite a match for the faster and more agile Japanese Zero made by Mitsubishi. O'Hare's squadron VF-3 became known as 'Fighting Three', renowned for its excellence. Its insignia – the popular cartoon figure Felix the Cat carrying a bomb – was one of the most famous in the US navy. Flying a plane off a carrier required special skills and nerves of steel. Aircraft were catapulted off the deck and when pilots returned to their warship, they had to land on a tiny runway bouncing around in the ocean.

By the time O'Hare graduated, Europe was entering year two of the Second World War. US President Franklin Roosevelt had kept his country out of the conflict so far but that was soon to change. On 7 December 1941, 353 Japanese warplanes took off from six Japanese carriers and launched a surprise attack on Hawaii's Pearl Harbor, home of the US navy. Eighteen warships were sunk or severely damaged and nearly 2,500 US sailors, marines and soldiers killed. It was a devastating blow and, since the Japanese were allied with Germany, it propelled the USA into the war. Roosevelt's navy had had one lucky escape: unusually, not one of its vital aircraft carriers, including O'Hare's *Saratoga*, happened to be in Pearl Harbor that disasterous day. The Japanese, as they would later learn to their cost, had missed the most important targets of all.

O'Hare knew his ship was safe. Like him, she was in San Diego, California. He heard the news about Pearl Harbor on the radio while driving home to the apartment he shared with Rita, his wife of three months. She remembered him bursting through the door shouting, 'We are to board *Saratoga* as soon as possible. It's war!'

Saratoga headed to sea but her luck ran out. On 11 January 1942 in mid-Pacific, O'Hare, now a lieutenant, was having dinner with fellow officers. Suddenly, a massive explosion rocked the vessel. One diner later recalled: 'It sounded like the bottom had been blown out of the ship.'

A Japanese torpedo had hit the port side, killing six sailors and causing serious damage to the ship. *Saratoga* needed a refit that would take months, so O'Hare and his air group joined her sister ship USS *Lexington*. She set off in a task force heading south to the Bismarck Archipelago in the South Pacific where Japanese troops had occupied the town of Rabaul, dangerously close to Australia. US Task Force Commander Vice-Admiral Wilson Brown decided to feign an attack on Rabaul in the hope of luring Japanese planes out to sea where his fighters and anti-aircraft guns could attack them. His ruse worked.

At 3.42 p.m. on 20 February *Lexington*'s radar operator picked up a formation of aircraft seventy-six miles due west of the task force. They disappeared for a few minutes then popped up again just forty-seven miles away and closing fast. The call 'torpedo defence stations' blasted through the ship. Then came: 'Pilots, man your planes.'

Butch O'Hare had been waiting for this command for weeks. He jumped into his Wildcat but was forced to sit patiently while six other pilots were catapulted off the *Lexington*'s deck to meet the first wave of the Japanese attack. They managed to destroy five of the attacking Mitsubishi G4Ms. The Americans jokingly dubbed these cigar-shaped torpedo bombers 'Bettys' after one airman's well-endowed girlfriend.

O'Hare was desperate to join the action, but by the time his plane took off he complained: 'There were so many of our fighters, I couldn't get in on the brawl.'

O'Hare's wingman Duff Dufilho flew next to him offering protective support. Dodging anti-aircraft fire from the American ships, the two pilots watched Japanese bombs falling dangerously close to the task force and wondered if their ship would be there to land on when they came down.

With the other US fighters preoccupied with chasing away the first wave of Japanese planes, O'Hare and Dufilho unexpectedly found themselves diverted to attack eight new bombers approaching from the other side. The Bettys had sighted the American warships and were heading straight for them. O'Hare and Dufilho were the only airmen between the attackers and the ships. They quickly tested their weapons but Dufilho realised that all four of his 0.5-calibre Browning guns were jammed and wouldn't fire. 'Go back to the ship,' signalled O'Hare, but Dufilho stuck with him, hoping to divert enemy fire.

The two pilots took their planes up to 12,000 feet, 1,000 feet above the enemy planes where they could not be seen. O'Hare knew he had two or three minutes to stop the bombers from dropping their loads on the American ships. He later said, 'There wasn't time to sit and wait for help. Those babies were coming on fast and had to be stopped.'

As the first Japanese bomber on the right side of the formation went past, O'Hare swooped down, got within 100 yards of it and started shooting. His bullets tore into the fuel tanks and the Betty lurched to the right and dropped out of the formation. Seconds later O'Hare hit the tanks of a second Betty and watched as it too left the formation streaming black smoke. At the time he thought he'd destroyed both planes but they managed to fly on. He later told a radio interviewer: 'When one would start burning, I'd haul out and wait for it to get out of the way. Then I'd go in and get another one. I didn't have time to watch them fall. When they drop out of formation, you don't bother with them any more. You go after the next.'

Crossing over to the left side of the enemy flight, he went after the rear bomber, hitting an engine and fuel tank. The aircraft fell away. O'Hare then targeted a plane in the centre and managed to get close enough to shoot and kill the pilot. A surviving Japanese pilot later gave this dramatic description of the dogfight: 'Enemy planes swirled round us, zooming, banking, diving, attacking us from all directions.' He had no idea that Butch O'Hare was the sole attacker.

On his third pass O'Hare knocked another Betty out of the air. That gave him a clear shot at the lead bomber. He hit the left wing causing an explosion that sent its engine tumbling into the ocean below. The plane itself soon plunged into the water. There were just two Bettys left. One more pass should be enough to dispatch them. But O'Hare's guns had run out of ammunition. The Bettys escaped but without doing any damage to the ships.

In just four minutes Butch O'Hare had destroyed four Japanese bombers and seriously damaged two more. He was convinced he'd downed all six. When he landed back on the *Lexington* and was asked how he was, he simply replied, 'I'm okay. Just load up those ammo belts, and I'll get back up.'

Later he described what it was like taking on an enemy formation single-handed: 'You don't have time to consider the odds against you. You are too busy weighing all the factors, time, speed, holding your fire till the right moment, shooting sparingly. You don't feel you are throwing bullets to keep alive. You just want to keep shooting.'

News of O'Hare's extraordinary flying feat swiftly went round the ship. When told he'd been recommended for a medal, his response was typically modest: 'I don't want a medal. The other officers in the squadron could have done the same thing and we all know it.'

The praise for O'Hare reached military bosses in Washington, the White House and eventually the American public. His exploit was a huge boost to national morale. Until now, the Japanese had seemed unbeatable but here was a victory that signalled the Pacific war could yet be won. While *Lexington* headed back to Pearl Harbor, newspapers were alight with patriotic fervour. The *Chicago Sun* called O'Hare 'an average American. Before they are through with us, the Axis and the Japs will find out that there are several million more where he came from.'

When Butch O'Hare stepped back on to American soil on 26 March 1942 he discovered, much to his embarrassment, that he had become a national hero. Uncomfortable in the limelight, he claimed he'd rather face Japanese bombers than reporters. He told interviewers he had done nothing unusual – that he was just doing the job he had been trained for. But that didn't stop the interest. The *New York Sun* waxed poetic:

> Japs? Where?
> Page O'Hare!
> Six! There!
> Up, O'Hare!
> None in air –
> Thanks, O'Hare.

On 21 April O'Hare and his wife were invited to the White House where President Roosevelt announced Butch had been promoted to lieutenant commander. He then presented him with the blue-ribboned Medal of Honour and read out a citation that mentioned O'Hare's valour in taking on so many Japanese planes: 'Without hesitation, alone and unaided, he repeatedly attacked this enemy formation, at close range in the face of intense combined machine-gun and cannon fire . . . As a result of his gallant action – he undoubtedly saved his carrier from serious damage.'

O'Hare spent the next few months helping the war effort. Other countries might have put him straight back into the cockpit to make maximum use of his killing power. But the American government decided he was more valuable mobilising support for the war. The popular hero promoted the sale of war bonds, appeared before a congressional committee to help lobby for increased naval funding, made motivational speeches to student airmen and starred in a homecoming parade in St Louis. More than 60,000 people turned out to see 'our Butch'. It was the largest crowd since the St Louis Cardinals won the 1926 National League baseball championship.

Finally, in mid-June 1942, O'Hare got back to doing what he really wanted, flying combat aircraft. Only this time he took command of his beloved squadron. He proved an excellent and popular leader, and he was soon flying America's latest fighter – the Grumman F6F Hellcat. This was a faster, but no less rugged 'big brother' to the

Wildcat. It had six guns instead of four and its final version could reach over 400 miles an hour. It would prove more than a match for the Japanese Zero. O'Hare taught new recruits how to get the most out of the new plane and how to survive a vicious dogfight. It was vital, he told them, 'to look behind you every chance you've got . . . even when you're pulling the trigger'.

By the time O'Hare went back to war the struggle with Japan had been transformed. Slowly but surely the tables were being turned and it was now the Japanese who were under serious pressure. He and his unit joined the USS *Independence*, a new light carrier, part of a task force attacking the Japanese-held Marcus Island, 1,000 miles south-east of Tokyo. On 31 August 1943 O'Hare led four Hellcats on a low-level strafing run over the island and then, almost touching the waves, finished off an enemy ship. His reward was a Distinguished Flying Cross 'for grim determination and courageous disregard for his own personal safety'.

A gold star was added to his Distinguished Flying Cross for actions in the battle for Wake Island, a key Japanese airbase 2,000 miles west of Hawaii. On 5 October O'Hare shot down his first Zero. It took him two passes to destroy a second plane because only one of his six guns was working.

A month later he was promoted to Commander of Carrier Air Group Six in the South Pacific. He joined the US carrier *Enterprise*, nicknamed the Big E. As he worked on a plan to destroy Japanese coastal defences on the Gilbert Islands he took time to reassure young pilots who had never been to war. One remembered O'Hare telling him that American fighter planes were better than anything the Japanese had: 'So you can't lose! Go in there and knock 'em out of the sky.'

On 19 November, after a traditional combat breakfast of steak and eggs, O'Hare led eighteen Hellcats on a raid that destroyed Japanese gun positions on Makin, one of the Gilbert Islands. He joined in strafing runs so low he nearly touched the tops of palm trees. His wingman noticed: 'Butch seemed in his glory . . . The flying risks and the in-the-air leading of his air group were his style of command.'

During the nine-day voyage to the Gilbert Islands, O'Hare was doing more than just planning an aerial attack on Makin. He invented a crafty way of allowing American fighter planes to engage in combat at night. Japanese pilots had already shown they could bomb ships under the cover of darkness. Since so many of their planes were being shot down by day, it was feared they would step up night bombing runs. The Americans had not yet sent night fighters from Pacific carriers. O'Hare was asked to change that. After consulting aviation experts on board, he came up with this plan: as soon as a ship's radar spotted enemy aircraft at night, a select team would take off. The lead plane would be a radar-equipped Grumman TBF Avenger. It would guide two Hellcats – which didn't have radar – towards the enemy planes, close enough for the pilots to see their blue exhaust flames and shoot them down. O'Hare named his new night fighters Black Panthers.

On 26 November O'Hare volunteered to go up at night to put his new technique to the test. It was the first time in history that American planes would take off on a combat mission from a carrier after dark. A colleague recalled: 'He was enthused about the new night tactics and seemed particularly anxious to be in the air that evening.'

It proved to be a busy but deadly evening. Fifteen Mitsubishi Bettys loaded with torpedoes were heading for the American ships. The announcement came at 5.45 p.m. while O'Hare was having dinner: 'Night fighters, man your planes.' O'Hare jumped up, telling the waitress to keep his ice cream cold until he got back. When a mechanic wished him good luck, O'Hare replied, 'Hell, we don't need luck with these cookies.'

Even as his Hellcat took off, a ship's radar picked up the approaching enemy bombers. O'Hare flew on the right, his wingman Ensign Andy Skon on his left. Skon recalled O'Hare asking him over the radio: 'Which side do you want, Andy? This was typically thoughtful considerate Butch.' Skon replied he'd take the port side, and O'Hare responded, 'Roger.' In front of them, leading the way through the darkness, Lieutenant Commander Phil Phillips piloted a radar-equipped Avenger. Anxious not to shoot at one of his own aircraft, O'Hare radioed Phillips: 'Phil, turn on your lights.' The rear-facing gunner in the Avenger's turret, Alvin Kernan, described seeing O'Hare clearly in the Hellcat's cockpit: 'Canopy back, goggles up,

Mae West . . . seated aggressively forward, riding the plane hard, looking like the tough Medal of Honour recipient, American ace he was, Butch O'Hare's face was sharply illuminated by his canopy light for one brief last moment.'

Suddenly Phillips spotted enemy aircraft and shouted, 'I have them in my sight. Attacking!' At 7.23 the Avenger shot down a Betty. Two minutes later came this radio message from O'Hare: 'My God. There's a tracer come through my wing. What a funny sensation.'

His plane apparently undamaged by the tracer bullet, O'Hare watched as Phillips in the Avenger shot down another Betty. Suddenly, yet another Japanese Betty appeared above the three American planes. Phillips warned O'Hare: 'Butch, this is Phil. There's a Jap on your tail,' and he told Kernan, his gunner, to open fire. The Betty closed in on O'Hare, firing its machine gun. Skon, on O'Hare's left, watched in horror as his commander's burning plane slid beneath him and disappeared from view. Skon and Phillips both radioed O'Hare: 'Butch, this is Phil. Over. Butch this is Phil. Over.' No reply. Phillips called the *Enterprise*: 'Butch may be down.'

The ship's log recorded that at 7.34 Butch was reported to have landed in the water twenty-six miles from the ship. Fear of submarine attack meant that a search party could not be launched until first light the next morning. It found pieces of enemy planes and an upside-down Japanese life raft. There was no sign of Butch O'Hare.

For years experts have debated whether O'Hare was brought down by an enemy aircraft or by friendly fire from the Avenger flying in front of his Wildcat. The general consensus is that he was shot by a Japanese gunner in the nose of the Betty flying above him.

Butch O'Hare's final medal – a Navy Cross – was awarded after his death. Its citation reads: 'Lieutenant Commander O'Hare's brilliant leadership and courageous initiative in the face of grave peril undoubtedly prevented the infliction of serious damage upon an important Task Group.'

Butch O'Hare's night flying initiative may indeed have saved many American lives but it cost him his own. The irony is that this hero's extraordinary career and premature death might never have happened were it not for Al Capone.

ESCAPES

FEW WARTIME WARRIORS can claim to have equalled the feat of Henri Giraud, a French officer who escaped from the Germans in both World Wars. In the First he dodged his German captors in a prison camp in Belgium and in the Second he shinned down the sheer cliff wall of Konigstein Prison on a makeshift 150-foot rope. Other escapes have stretched human ingenuity and stamina to astonishing limits: they've been celebrated in films such as *The Wooden Horse*, the story of the men who dug a tunnel out of a Nazi prison camp perching inside a wooden vaulting horse. It was placed on the same spot every day by a group of British inmates pretending to be gym enthusiasts. The Germans never discovered that men hidden inside were digging a tunnel beneath it and removing the sand by hanging their bags of spoil inside the horse as it was carried to and from their prison hut.

There's a whole stack of remarkable escape stories in which prisoners of war, desperate to evade the boredom and brutality of wartime jails, risked their lives to return to the freedom of their own countries. We've chosen two: the first is the extraordinary tale of a captured German pilot whose audacity outdoes that of any other escapee in the Second World War. He was one of very few Germans in the whole war who escaped from the Allies and made it back to Germany. The other is Eric Newby whose chronicle of his flight from the Fascists in Italy is outstanding for its melodrama – and for its humour.

But we've also taken one story from a multitude of harrowing accounts by refugees who've escaped from the most destructive conflict so far of the twenty-first century – in the Middle East, where the single most war-torn country has been Syria. More than 10 million, nearly half its population, have fled their homes, and some 5 million have sought shelter abroad. All of them tell anguished stories. We were so moved by that of Ahmad Terkawi, his wife and their two small children that we travelled to Sweden to meet them. But first – one of the coolest escapers of them all – Franz von Werra.

26

Franz von Werra

Luftwaffe escapee in UK and Canada, 1940–1

FRANZ VON WERRA, we must admit, was a bit of a show-off. He had a reputation among his fellow Luftwaffe pilots for bumptiousness and self-promotion. He mastered the art of flying easily and was, by September 1940, as agile at manoeuvring his Messerschmitt 109 as most of Germany's Second World War aces. His skills won him the Iron Cross but his claim of nine kills of RAF planes in one day was met with laughter by his peers. He said he'd destroyed four in the air and five on the ground. It was a preposterous claim and he was only allowed to chalk up five notches on his fuselage. But we forgive him his boasting because he has presented us with the most extraordinary escape story.

On 5 September 1940, two months into Germany's battle to break the Royal Air Force and pave the way for a Wehrmacht invasion of Britain, Von Werra's plane was escorting a cluster of German bombers 6,000 feet below. They had just attacked Croydon and were turning for home in an almost cloudless sky. His radio crackled urgently even before he saw the three Spitfires diving past him to attack the bombers. '*Achtung! Achtung!* Attack!' came the crisp order from the commander of his *Gruppe*. Von Werra swung to his left and dipped his nose. He was too late. He suddenly saw another Spitfire behind him. It was firing and his aircraft couldn't dodge the bullets. Within seconds his engine was stuttering and he was losing control. He prepared for a crash-landing.

Residents of Kent had become inured to the dreadful racket in their skies almost every day that summer. But this time it wasn't just noise that grabbed the attention of a battery of anti-aircraft gunners on a farm at Marden ten miles south of Maidstone. They watched a crippled ME 109 dip low over trees and crunch into the ground. The gunners saw the pilot jump out rapidly and set fire to his papers. They saw him look around desperately for an escape route but when they raced towards him, their guns at the ready, he made no attempt to resist.

Von Werra was soon in a police cell in Maidstone under the watchful eye of Police Sergeant Harrington who found him 'confident, alert, highly intelligent', and, oddly in the circumstances, with a keen sense of humour. The German prisoner spoke fluent English. He told Harrington with a laugh that the RAF had prevented him keeping an important dinner engagement that night. 'He struck me as being a bit conceited though,' Harrington reported. 'When I asked him his rank he clicked his heels together and said he was an Oberleutnant and a Baron.' He flashed his big signet ring at Harrington and boasted that his father owned several castles in Switzerland. He was half right. He was indeed born to a Swiss family that owned several castles but his aristocratic father had gone bankrupt and sold them.

Taken by truck to London, von Werra got a glimpse of how well Britain had prepared for a Nazi invasion. There were roadblocks, tank-traps and pillboxes on main and side roads to left and right. The war was a year old. The British army, defeated on the Continent, had managed to extricate itself at Dunkirk, but Britain was the only European power left fighting Hitler. The Battle of Britain between the RAF and the Luftwaffe for air supremacy would decide whether Hitler could invade. But now von Werra was grounded. He cursed his bad luck and imagined his fellow pilots climbing out of their warplanes back in Germany wondering what had happened to him.

In London von Werra was given a medical examination: 'Age 26, height: 5ft 7 ins, weight: 10 stone, appearance: sturdy build, fair, wavy hair, blue eyes'. He was issued clothes and told their cost would be deducted from his first month's allowance. His rank entitled him to £3 a month. He then underwent a rigorous interrogation by a British squadron leader who ridiculed him for claiming 'phantom' kills of British aircraft. Von Werra riposted that he'd hardly expect the British to admit to such losses. He then went on to bet his interrogator he would escape within six months. During several days of questioning, von Werra told the British nothing, but he learned a lot about British interrogation methods.

At the end of September he was moved to Grizedale Hall prison camp in wild moorland in the Lake District 250 miles north-west of London. The hall, since demolished in 1957, was a large and gloomy stately home built by a wealthy shipping magnate in 1905 in the neo-Gothic style. It was nicknamed the 'U-boat Hotel' as many of its inmates were captured German submariners. There were Luftwaffe captives too including Major Willibald Fanelsa, the senior German prisoner of war at Grizedale. He was also the leader of the prisoners' council – the *Ältestenrat* – which vetted all escape plans. He and his colleagues were surprised to be confronted by a cocky von Werra only ten days after his arrival. 'I have a plan for escape,' he told them. During closely supervised exercise walks outside the prison he had noticed a spot where he could drop over a wall and into a meadow without being spotted by the guards. Using the tallest prisoners as a screen he planned to lie flat on the wall and: 'as soon as the guards' attention has been momentarily distracted by four other prisoners, a nudge will give me the signal to drop off the wall and run behind it along the edge of the meadow and into a wood.' He intended to head for the west coast and find a neutral ship going to Ireland.

The council agreed to support him, although one U-boat commander, who'd made an abortive attempt to escape earlier, warned him the country was 'extremely hard going . . . it's a switchback of hills and dales'. Fanelsa told von Werra his escape would trigger an immediate response from two battalions of troops stationed in the area.

On 7 October 1940 twenty-four German officers including von Werra were escorted out of Grizedale Hall on their daily route march. It usually took place before lunch but Fanelsa had persuaded the British to move the walk to the afternoon. It would be easier to escape as darkness set in. They reached a corner beyond the village of Satterthwaite and stopped as usual to rest against the wall von Werra had spotted. With eight prisoners standing around him, and the added distraction of a passing horse and cart, he rolled off the wall and hid behind it. Two women working in the field saw him and waved and shouted to the guards but they were out of earshot. The remaining Germans waved cheerily back, neatly deceiving the British guards into thinking the prisoners were looking for female company. It was a nervous moment for von Werra but once his comrades were marched off by the unsuspecting British, he too waved gaily at the women and sprinted into a wood.

The British soon realised there had been an escape. While the delighted Germans celebrated that evening with songs and shouts of *Sieg Heil!*, a huge manhunt was mounted, unsuccessfully. There was no sign of von Werra. For three days as rain bucketed down the fugitive struggled through the mud and undergrowth towards the sea. He had a narrow escape when two members of the Home Guard found him sheltering in a hut near Barrow-in-Furness. They tried to arrest him but he shook himself free and outran them. It was another three days before he was next sighted. Utterly exhausted, he knew he'd been seen and threw himself face down in a muddy hollow. The sheep farmer who spotted him alerted a nearby search party. There was no escape this time and von Werra was soon in handcuffs. All he had on him was a table knife, a bar of chocolate, his shaving kit in a tin and his Iron Cross. Back at Grizedale he was treated to twenty-one days in the 'cooler', the solitary confinement cell.

On 3 November he was transported to another prison camp, Swanwick, near Derby. It was even more secure than Grizedale but von Werra was soon plotting his next escape. Major Fanelsa had also been switched from Grizedale and was now senior officer at Swanwick. He made it clear that the last thing he wanted was another bid for freedom by von Werra, which might upset his own comfortable relationship with the camp commandant. So von Werra looked elsewhere for support and was delighted to find an enthusiastic group of Luftwaffe officers who applauded his plan to tunnel a way out. They laughingly dubbed themselves the Swanwick Construction Company. Thirteen metres of burrowing from under the floor in their block would take them beyond the wire. The earth could be dumped into a huge rainwater tank.

They started digging on 17 November. After a few metres they ran into a formidable obstacle – a huge lump of sandstone. They could smash their way through it with a crowbar but it would be noisy, so all sorts of loud games, quarrels, even fights had to be faked in order to distract the guards. The next blockage was a hefty sewer pipe. They couldn't tunnel under it, because water welled up, so they dug over it. It was so close to the surface that the roof fell in and von Werra and his exhausted fellow digger were lucky to get out alive.

But somehow they managed to keep burrowing and at last the tunnel was ready. There were five in the escape team. Four would try to make it to the coast and look for a neutral ship. But von Werra's experience at Grizedale had persuaded him that only a quick getaway would avoid capture. That meant escape by air and von Werra – typically – dreamed up the most brazen of plans. He would head for the nearest RAF airfield, claim he was a Dutch pilot whose RAF plane had crashed and demand another aircraft to fly back to his base in Scotland. Once off the ground he would head straight home to Germany.

On 20 December, nearly the shortest day of the year, the five men emerged from their tunnel and made their escape in the pitch dark. To distract the guards other prisoners – including even the spineless Major Fanelsa – struck up a chorus of folk singing. There was a heart-stopping moment when three guards approached the tunnel's exit, but at the last second they turned away. The prisoners were free. If all went well they would have ten hours before the morning roll-call at Swanwick triggered a massive chase. Von Werra made his own way across the fields in search of an RAF station. He had scavenged a flying suit from one of the German pilot prisoners. Another had given him a fine pair of boots. He had no idea where the nearest RAF station was but happened to stumble across a railway line before dawn where a steam engine was stationary in a siding.

As he approached, the driver stuck his head out of the cab and asked what he was doing on the railway line.

'I am Captain van Lott, formerly of the Dutch Royal Air Force, now serving with the RAF. I've just made a forced landing after being hit by flak in a raid over Denmark. I must get to the nearest RAF aerodrome as quickly as possible.' The engine driver's first instinct was to laugh but the man in the flying jacket looked convincing enough and the urgency in his voice demanded action. His accent could well be Dutch. Within minutes von Werra was at Codnor Park railway station where he met the booking clerk just after 6 a.m. He explained that his Wellington bomber had crashed two miles up the line and he now had to get back to his unit in Scotland. Would the booking clerk please telephone the nearest RAF station and get a car to collect him? The clerk picked up the phone but to von Werra's alarm he called the police station.

'Don't worry,' said the railwayman reassuringly. 'The sergeant said someone would come right away.' Von Werra thought of making a run for it but he knew that would be a fatal mistake. Better to play along. There was no sign of the police after twenty minutes and he decided to up his game. He told the clerk that the reason for his urgency was that he'd been testing a secret new bombsight and it was essential he return to his base in Aberdeen immediately. The clerk, clearly embarrassed at the delay, picked up the phone and got through to RAF Hucknall nearby. He handed the receiver to von Werra who then explained his predicament so urgently that the RAF duty officer agreed to send him a car.

Just as von Werra was congratulating himself, in walked three policemen looking very grim. Little did von Werra know it but one of his fellow escapers had made the mistake of stealing a policeman's bicycle and promptly been caught. But von Werra's fluent English and the derring-do of his story about the raid on Denmark and narrow escape in the crash worked their magic on the police officers. They wished him luck and soon the RAF driver turned up and swept him off to see the duty officer at Hucknall.

Von Werra couldn't believe it when RAF Hucknall's gate was opened for him without any questions and he spotted a group of Hurricane fighters lined up by the control tower. In an hour or so he could be soaring away over the Channel coast and across to Germany. 'Don't bother to go any further,' he said obligingly to the driver. 'Just drop me at the Control Tower.'

'Sorry, sir, but I have to take you to the duty officer,' replied the driver. Von Werra found himself facing a squadron leader who was suspicious enough of this bizarre fellow's telephone call to take the precaution of locking all the building's windows and doors except the entrance. But the charm of this personable pilot soon worked on him. The squadron leader was impressed by his boots and flying jacket, which von Werra said he'd been given by the Dutch airline KLM when he'd flown for them before the war.

When asked for his RAF identity card, he replied that on special missions aircrew did not carry their papers. The squadron leader seemed to buy von Werra's story but said that just to check he would telephone RAF Dyce, the Scottish airbase his visitor said he came from. Fortunately for von Werra the call took a long time to connect. He signed to the squadron leader that he was going to visit the toilet, which he had noticed opposite the entrance door. While the duty officer struggled with his phone call, von Werra raced out the door and headed for the Hurricanes.

He strode up to an engineer servicing one aircraft and announced he'd been cleared for a practice flight. Would the man be kind enough to show him how to use the controls? The mechanic replied that he would have to sign the necessary papers. So they went to a nearby office where von Werra scribbled a fake signature and then hurried the engineer back to the aircraft. He expected the duty officer to send RAF police after him at any moment but the poor man was still fighting to get through to Scotland. Von Werra was soon in the Hurricane's cockpit listening intently as the engineer explained the controls. Once the man was out of the cockpit von Werra pushed the button marked 'Starter'. 'Don't do that,' shouted the engineer. 'The engine won't start without a starter trolley.' He was beginning to wonder about von Werra, but that well-practised charm worked again; the trolley was hauled up and the cable plugged into the Hurricane. All von Werra had to do was to press the starter again. But suddenly he felt the barrel of a pistol against his temple. He turned and found himself looking into the muzzle.

'Get out!' snapped the duty officer. Von Werra's second escape had so nearly succeeded. But – as his captor smugly put it: 'I finally got through to Dyce.'

A few people at RAF Hucknall later expressed their admiration for von Werra's chutzpah. One tantalising detail about his last-minute failure is that the Hurricane he tried to steal was a new experimental version. It would have been a magnificent prize for the Luftwaffe.

British military chiefs must have heaved a sigh of relief when it was decided to move some German prisoners of war in Britain to Canada. They were particularly delighted to see von Werra packed off aboard the *Duchess of York* on 10 January 1941. Needless to say, he immediately began plotting another escape. One of his hare-brained ideas was that he and his fellow prisoners should seize the ship. But since the convoy was escorted by the battleship HMS *Ramillies* the whole way to Halifax, Nova Scotia, he reluctantly dropped that plan.

The long train journey taking the prisoners nearly halfway across Canada to a prisoner-of-war camp was another matter. Von Werra quickly spotted that the railway ran close to the border with the United States, still neutral in 1941. If he could jump off the train beyond Montreal where it passed within reach of the St Lawrence River he could find a way to cross it into America. Von Werra took a seat by a window. It was frozen shut in January but he noticed that the heat inside the carriage reduced the frost on the glass. Von Werra managed to wrench it open a fraction and shove a small piece of paper under it to keep it ajar. His fellow prisoners obligingly prepared to stage a diversion. One got a blanket ready to conceal his escape. After Montreal the train seemed to rattle along for miles without stopping: von Werra had to wait for it to slow down. Near Smith's Falls, Ontario, the train halted and then slowly moved off again. Here was von Werra's chance. On a sign from him, other prisoners rose to their feet and crowded into the aisle. With the blanket screening him, von Werra flung open the window and threw himself out. He landed in a huge pile of snow.

Other than a few bruises and stiff neck, he was undamaged. It was intensely cold and he had to keep moving. The sky was clear and so he put the North Star behind him and trudged south through deep snow. He walked thirty miles to the north bank of the St Lawrence, and, finding a road, hitched a lift to the Canadian town of Johnstown opposite the US city of Ogdensburg. There's a bridge there today, but in 1941 the river was a 1,000-metre-wide expanse of what looked like solid ice. Freedom was only a walk away but the snow on the surface was deep, the wind howling and the chill agonising. As he trudged through the snow and ice, lights on the other side came closer, but then he heard the sound of rushing water. Suddenly he was on the edge of a black torrent where the river hadn't frozen over. Von Werra knew that to risk swimming would be suicide. He turned around and trekked back to the north bank utterly exhausted; however, he wasn't going to be cheated of this third chance of freedom. On the riverbank he found an upturned rowing boat. He dragged it – only a metre or two at a time – across the ice and snow until he reached the open water, heaved it in and drifted downstream. The current was fast and he was lucky to strike the opposite bank and lodge there just long enough to stagger on to the shore. A road ran along the river and he spotted a man and a woman stooped over the open bonnet of their stalled car. To von Werra's massive relief it had New York number plates. Minutes later the exhausted German surrendered with delight to an American policeman.

Canada demanded his return but Germany insisted he was on neutral US territory. A frantic legal battle ensued but von Werra once again slipped the net. With the help of German consular officials he crossed the Mexican border. From there he made his way to South America. He flew first to Lima and then on to Rio de Janeiro, across the Atlantic to Italy and finally home to Germany, arriving on 18 April 1941.

There was spectacular press coverage in Germany of his escape to the USA and von Werra received a hero's welcome when he reached home. He was decorated by Hitler himself. Hermann Goering, head of Germany's air force, promoted him to Hauptmann, the equivalent of an RAF flight lieutenant. Von Werra wrote a book about his adventures but he was so complimentary about his British captors that publication was banned by Hitler's propaganda minister, Joseph Goebbels.

Months later, in June 1941, Germany invaded Russia and von Werra rejoined the Luftwaffe. After a short time on the Russian front, he was moved to Holland. Goering was not going to risk him flying over Britain again, so he was assigned to fly a brand-new Messerschmitt 109 in a defensive role along the Dutch coast. By now von Werra had married and he and his new wife, Elfriede, rented a fisherman's cottage. Von Werra's nephew, Hans, was eight years old at the time and living in Holland. He told us how his uncle, 'my famous Uncle Franz', came to the family home in The Hague where a party was planned for the Luftwaffe hero three days later. But on the night of the party, 25 October, two German officers in long black coats knocked on the door. Franz, they told the family, would not be coming to the party. Earlier that day he had flown his aircraft over the cottage he'd rented with his wife of two months. He dipped his wings to salute her. Moments later as he headed out to sea his engine developed a fault and his aircraft was seen plunging into the waves. In spite of a painstaking search ordered by Goering himself no trace of von Werra or his aircraft was ever found. It was only thirteen months after he'd crash-landed in Kent and been taken prisoner by the British.

27

Eric Newby

Escape in Italy, the Fugitive with a Sense of Humour, 1943

> We were captured off the east coast of Sicily on the morning of the twelfth of August, 1942, about four miles out in the Bay of Catania. It was a beautiful morning. As the sun rose I could see Etna, a truncated cone with a plume of smoke over it like the quill of a pen stuck in a pewter inkpot, rising out of the haze to the north of where I was treading water.

ERIC NEWBY'S LIGHT-HEARTED account of one of the worst moments of his life, when his boat sank after a raid on Sicily, is typical of his beguiling style. From its very first line, Newby's eagle-eyed powers of observation, lyrical writing and jaunty sense of humour offer a refreshingly different take on a great wartime escape.

From the moment Italian fishermen picked him up and he was made a prisoner of war, he resolved to get away. And he did – spending sixteen extraordinary weeks on the run before being recaptured by the Fascists. Incarcerated in German and Czech prisoner-of-war camps for the last years of the Second World War, Newby came up with the perfect pastime: he spent his days jotting down memories of his daring escape into the mountains of Italy. A quarter of a century later he turned his notes about this remarkable experience into a compelling book. We happened to meet Eric Newby and his wife Wanda in 1986 and they regaled us with funny stories they were about to include in a book about a recent bicycle trip around Ireland. We had always enjoyed his marvellous travel writing but we then had no idea that the Newbys were war heroes.

A middle-class London lad, Eric Newby joined the Royal Navy's Special Boat Section – the naval equivalent of the Special Air Service – in 1941, aged twenty-two. His first major assignment seemed doomed from the start. Even its name was ridiculous: Operation Whynot. Without any proper surveillance photos, Newby and four others were expected to attack a German airfield in Sicily. The airbase, with its command over the Mediterranean, was operated by Hitler's Luftwaffe in cooperation with their Italian allies. Whynot's goal was to destroy up to sixty Junker JU 88s, which were extremely versatile German fighter planes. They were used to devastating effect against Allied ships supplying the strategically important island of Malta. It was from Malta that Britain's navy and air force could attack General Rommel's German army in North Africa.

The Whynot team left Malta in a submarine and surfaced in the dark thirty-six hours later near the Sicilian airfield. Newby headed towards a beach in a leaky rubber canoe. In his book, he remembers that as enemy aircraft came in to land just above his head, he desperately tried to cheer himself up by thinking of 'the bacon and eggs which we had been promised if and when we returned on board'. Swimming the final few yards to shore, the invaders primed their pencil bombs, which looked like 'big, black conkers on strings', buried their boats in the sand and cut through wire fencing surrounding the airbase. As they crept through a field, they were terrified by a noise that sounded like soldiers marching. Newby says he was relieved when it turned out to be a horse munching grass. Once past 'this nice enemy horse', they found themselves in a building full of aircraft engines. They stuck their pencil bombs into as many engines as possible and then moved on to tackle the planes. But they were spotted by an Italian patrol. Shots were exchanged in the dark and 'pandemonium broke loose'. It was a shambles. A man shouting in German loomed up in front of Newby, who raised his gun and was about to shoot when a Scottish voice yelled, 'Don't shoot, you stupid bastard. It's me.' Newby had nearly taken out his commanding officer.

Somehow team Whynot managed to run off into nearby woods which, Newby writes, reminded him 'vividly of Act Two, Scene One of *As You Like It*, the Forest of Arden in the Open Air Theatre in Regent's Park'. He continues the theme of Shakespearean farce when he describes one of his party falling through a groundsheet covering a hole and landing in the middle of a group of Italian soldiers playing cards. The man was promptly arrested but the rest made it back to the beach. There was a brief moment of joy when they heard the bombs they'd planted exploding. That was quickly replaced by panic as they scrambled through the sand searching for their canoes. When they finally found them it was already 12.45 a.m. They paddled frantically towards the submarine which was scheduled to depart at 1 a.m. They couldn't find it in the dark and were left tossing on an increasingly rough sea. Newby's boat filled with water and sank. He clung on to the single canoe still afloat.

After five hours in the water, he and the others were spotted by Italian fishermen 'who hauled us into their boat like a lot of half-dead fish'. He remembers lying among fish 'all displaying considerably greater liveliness than we did'. The fishermen took their human catch to Catania where they were interrogated by Italian officers furious that their defences had been breached. Told he would be shot at dawn the following day, Newby was tormented by stories he'd read at school about final moments of First World War spies. 'Mostly they were cowardly spies whose legs gave way under them, so that they had to be carried, shrieking, to the place of execution and tied to stakes to prevent them sinking to the ground and, although I hoped that I wouldn't be like this, I wondered if I would be.'

Fortunately he was never put to the test. The Germans, angrier about the Italians' decision to shoot the British without consulting them than about the attack on the airfield, ordered the prisoners taken to Rome for further questioning. Here Newby met real traitors for the first time. His Irish and South African interrogators were soldiers who'd changed sides after being captured. They called him 'old boy' and offered him 'nights on the town and escape routes into the Vatican in exchange for information'.

Newby refused their enticements and ended up in a prisoner-of-war camp for officers at Fontanellato near Parma in the Po Valley. He amused himself by observing his fellow inmates, noting that upper-class officers were reluctant to socialise with other prisoners. But there were not enough posh men to fill a twenty-seven-bed dormitory so they were forced to accept people like him: 'marginally OK people, the sort of people they were prepared to talk to and drink with while the war was on, and then would never see again'.

'Marginally OK' Newby watched fascinated as 'Bolo' Bastonby and 'Feathers' Farthingdale played baccarat just as they had done at their London clubs before the war, then wrote home to their banks for money to cover their losses. The prisoners enjoyed ogling local girls who regularly strolled along the road next to the camp. The best view was from a room that doubled as a bar at the top of the building. Since it was forbidden to look out of windows, watchtower guards took great delight in firing shots at the peeping Brits. No one was hit but Newby reports that the bullet holes in the bar walls 'gave the place a raffish appearance, like a middle-western saloon built by some renegade, gun-toting priest'.

On 8 September 1943, thirteen months after his capture, Newby was lying in bed in a medical room. He'd broken his ankle running up and down the camp's marble staircase to keep fit. Suddenly shouts of *Armistizio!* filled the air. The Italian government had agreed an armistice with Allied Commander General Dwight D. Eisenhower and all hostilities between the two sides had ceased. The Italians caved in only five days after Allied forces landed on the Italian mainland but it was to take another eighteen months for the German army in Italy to be defeated.

Knowing that German soldiers would soon arrive to take over the prisoner-of-war camp, the British prisoners seized this chance to escape. Unable to walk on his broken ankle, Newby was carried to a horse named Mora. He was terrified. He'd never ridden a horse before, and Mora was particularly stubborn. She stopped at the first ditch and refused to go further until someone 'stuck a lighted cigarette up her chuff'. Mora bucked angrily, throwing Newby over her head. In spite of landing on his bad foot, he still had the wit to note that everyone 'was cheered up by this spectacular happening'. Carried to a farm and left to fend for himself, he spent the night in a hayloft – the noise of passing German trucks punctuated by his own sneezes. He wished he was back in England where 'the hay fever season would have been over'. The next morning a doctor appeared, 'an enormous, shambling man with grizzled hair, like a bear and one of the ugliest men I had seen for some time'. He insisted on taking Newby to a hospital. As he got into the doctor's tiny Fiat, Newby noticed 'a bold, good-looking girl' with blonde hair and blue eyes. She smiled at him and promised to visit him in hospital. Little did he know he had just met the love of his life.

The hospital was next door to the prison camp he'd just escaped from but Newby didn't mind. Compared to the rest of his war, this was heaven. A cast was put on his foot, nursing nuns brought him delicious food and he fell asleep 'in heavenly clean sheets like a great cosseted baby'. Best of all, the beautiful blonde turned up and offered to teach him Italian. Her name was Wanda Skof, a Slovene whose father was a schoolteacher in Italy. Newby quickly fell for Wanda. He managed to grab a kiss out of sight of eagle-eyed nursing sisters but couldn't resist making fun of her fractured English, especially when she pronounced his name Eric as 'Hurruck'.

Their budding romance was rudely interrupted when, ten days after his escape, the Germans discovered Newby was in the hospital. Two guards were posted outside his door but he came up with an ingenious plan to get past them. He asked to go to the toilet. The guards said no, it was forbidden. When he announced he had diarrhoea, it had the desired effect: 'Like certain coprophilous German soldiers to whom the mere mention of excreta and, or bottoms, was sufficient to lay them on their backs helpless with mirth, they thought this very funny; but they let me go.'

The toilet and its window offered an escape route that proved a godsend when a nurse smuggled in a note from Wanda. 'Get out,' it read. 'Tonight. 22:00, if not, Germany tomorrow.' Determined not to end up in a prisoner-of-war camp in Germany, Newby acted swiftly. He told his guards yet again that he had a stomach ache. They'd now heard it so often they ignored him. He walked past them to the toilet, locked the door, opened the window and slid down a drainpipe. In pyjamas and bare feet, he limped through a bog to the spot where Wanda's note said her father and the doctor would be waiting. They shoved him into a car and drove through narrow, winding lanes to a wood where Newby was given a jacket, trousers, a blanket and some mosquito repellent. He was told a man called Giovanni would come and get him the next day. Exhausted, Newby lay down, lathered himself with repellent and went to sleep.

Giovanni, a Communist fisherman who hated the Nazis, turned up as promised and took Newby home. Germans were searching nearby houses for escaped Allied soldiers so Giovanni dug a grave-shaped hole where the fugitive, covered by a huge piece of turf, spent an uncomfortable night. The next day the doctor reappeared saying, 'This place is swarming with Germans. It's like Potsdam.' They drove off, passing an entire German Panzer Division on the road. Then in the city of Parma the doctor's car broke down. More German tanks rolled past and Newby was convinced the game was up. But at last the car started again and he was on his way to the inaccessible Apennine Mountains. He was dropped at the house of a farmer who had promised to hide him. Newby was just settling in when he was asked to leave. His host had changed his mind. When Newby asked why, the man replied, 'I am afraid of the Germans. I am afraid of the Fascists. I am afraid of my neighbours, and I am afraid of having my house burned over my head and of being shot if you are found here.'

The faint-hearted farmer sent Newby up a mountain path to a family called Zanoni, who lived in a tiny house built into a rock. The desperate fugitive asked if he could sleep in the hayloft. 'No you can't,' came the reply, 'you can sleep in my house in a bed, and you will too.'

Ushered into his generous host's own bedroom, Newby was invited to stay as long as he liked. But he was warned to keep out of sight because Italian Fascists, still loyal to Germany, were offering large rewards for captured Allied prisoners of war. Recognising that his presence was a danger to Zanoni, his wife and two children, Newby begged to be moved to an even more remote location. He was taken further up the mountain to another farmer named Luigi who took him in on condition that he worked for his keep. Work, Newby soon discovered, was twelve-hour days of back-breaking labour, clearing fields of huge boulders.

Of all his experiences Newby's relationship with Luigi and his family was the most bizarre. The day began with the banshee shriek of Luigi's wife, Agata: 'Time to get up.' It was a sound that 'never failed to make me leap from my bed in terror, believing the Germans had come'. The rock clearing was mind-bogglingly boring but, after stacking up the stones in an ancient cart Newby discovered he quite enjoyed disposing of them by throwing them over a cliff. Once he lost control of his cartful of rocks and nearly disappeared with it over the cliff. Luigi's only concern was for the cart: 'You watch that cart. It's a good cart. If that goes you'll have to carry them down in your hands.'

Newby's other great challenge was surviving the family dog, Nero. He had to pass the vicious animal to reach the outdoor lavatory which made him feel like 'a British ship heading to Malta being dive bombed by a German plane'. He adopted a strategy of moving very quickly and carrying stones as weapons. When the dog, foaming at the mouth, made a rush to intercept him:

> I altered course to port and rushed through what was the equivalent of the Sicilian Channel, the narrows between the house and a pigsty, jumping over his chain as I did so, at the same time raising one of the stones above my head in a threatening manner and roaring at him at the top of my voice, which sufficiently impressed him with my murderous intentions to halt him long enough to let me get through and out of biting distance.

When he wasn't outwitting Nero or heaving rocks, Newby entertained Luigi's household with stories about British murderers. Their favourite was Jack the Ripper. One month after Newby arrived at Luigi's someone informed on him and German soldiers raided the farmhouse. Fortunately he was not at home and so narrowly avoided arrest. Newby had to move again. Luigi's typically blunt words of farewell were: 'Lucky you finished the stones before all this happened. Otherwise, we would have to have done it ourselves.'

It was now too risky for partisans to take the man they called Enrico into their homes, so they built him a mountain hideout. On 27 October 1943, less than two months after going on the run, Newby moved into his new residence. It was in a cleft between two cliffs. Tree trunks supported a moss-covered roof, the front door hidden under the roots of a beech tree. Food was delivered daily, often by children. One day they also brought a message that made his heart leap for joy. Wanda was on her way to meet him. But the news she had for him was bleak. The doctor who'd driven him to safety had been arrested by the Gestapo. So had her father. (Both survived the war.) She also reported the Allied advance through Italy had bogged down between Naples and Rome which meant liberation was a long way off. Before they parted, Wanda gave Newby a parcel containing a compass, maps, a watch, blanket and Volume II of Gibbon's *Decline and Fall of the Roman Empire* which he had left in the prison camp. As she disappeared into the night, he remembered feeling incredibly lonely. He started crying, 'something I had not done for as long as I could remember'.

Newby was soon cheered up by another visitor – James, a fellow British officer and an old friend from the Fontanellato prison camp. He too had been in hiding, helped by brave Italians. A partisan had brought him to Newby and the two men settled in together. As it rained, then snowed outside, they spent twenty-two days in their sheltered hideaway reading and chatting about what they would do after the war. James would be a farmer. Newby dreamed of writing and opening a second-hand bookshop.

Then came the news that another informer had told the Fascist militia where they were. They hastily moved to a deserted barn. A local family, unfazed by threats of reprisals for harbouring Allied soldiers, invited them for Christmas. When their hostess asked what they would like more than anything, the two fugitives exclaimed in unison 'a bath'. And that's exactly what they got, in an old wine barrel. 'We stripped off in turns – it was no time for false modesty – and because the barrel was too close a fit for either of us to move our arms, she and her husband took turns to scrub us and wash our hair.'

They also had the extraordinary experience of listening on an illegal radio to King George VI's Christmas message. When he said, 'Some of you may hear me in your aircraft, in the jungles of the Pacific or on the Italian peaks,' Newby writes: 'The people in the room witnessed the awful spectacle, something which they are unlikely ever to see again, of two Englishmen with tears running down their cheeks.'

Eric Newby's escape, which began on 8 September 1943, ended abruptly on 29 December when, in his words, 'a dozen very nasty-looking men armed to the teeth with carbines and Beretta submachine guns' appeared at the barn. The Fascist militia had finally got their man. And James too. They spent the rest of the war in Nazi prison camps.

'Dad loved action and adventure,' Newby's daughter Sonia told us. 'He loved camping when he was a boy scout.' That helped him survive his time as a fugitive. But, typically, Newby made the best of being a prisoner of war. He told Sonia he hated being locked up but since he had not gone to university, he was delighted to discover that he was surrounded by knowledgeable fellow prisoners who gave regular lectures. So he seized the chance to learn.

After the war ended Newby returned to Italy working for MI9, a British agency that traced and rewarded partisans who had helped British soldiers. He went back to Fontanellato, the place where, two years earlier, he had made his clownish escape on the back of a horse. And there was Wanda, working in a bank. It was a dream come true. They were married in 1946.

Newby went on to become one of the most celebrated travel writers of his generation. The affectionate tale he tells in *Love and War in the Apennines* about the Italians who risked their lives to help the young British fugitive is one of our favourite pieces of literature.

28

Ahmad Terkawi

Escape from Syria, 2012–15

IN 2010 AHMAD Terkawi was a successful pharmacist in the Syrian city of Homs. He was twenty-eight, married to a charming young radiology student. They had a baby boy and lived comfortably in the suburb of Baba Amr where Ahmad's pharmacy did lively business. (We agreed not to name any family members except Ahmad.) The Terkawis looked forward to a future full of promise. Homs was a thriving commercial centre, Syria's third largest city halfway between the other two great cities, Damascus and Aleppo. It had a proud past. It prospered under the Romans and after the seventh-century Arab conquest it was a source of spirited opposition to the Abbasid rulers who moved the capital of the Arab world from Damascus to Baghdad. That rebellious spirit flared again 1,200 years later in Homs when tens of thousands of people demonstrated in the main square against the dictatorial regime of President Bashar al-Assad. Sixty-two people died in fighting between Assad's government and protesters in the city in the middle of April 2011. Homs was dubbed the 'Capital of the Revolution'. It wasn't long before it was being torn apart by a civil war that swept away all the hopes and expectations of families like the Terkawis and would soon engulf the whole country.

The young pharmacist was caught in the ghastly daily crossfire between government and rebels. Like most Syrians he had taken no part in political life. President Bashar al-Assad and his father Hafez had imposed a repressive stability on the country since the 1960s. And when the bloodshed began Terkawi took no side. Amid all the fighting, his pharmacy was destroyed by a rocket-propelled grenade. He is not sure whether the weapon that wrecked his business was fired by rebel or government forces. He had no time to investigate because he was immediately immersed in a humanitarian emergency that demanded all his energy and resources. He became part of a medical team fighting to save the lives of dozens of wounded men, women and children in Homs. They took over a store, put mattresses on the floor and set up a rudimentary medical centre. He and seven doctors treated scores of people. 'I was kneeling beside people dreadfully wounded,' he told us, 'tearing my shirt up to make bandages, and stitching wounds. There was nothing we could do for many casualties. They would come in with no legs. There were even headless bodies.' He treated everyone who was brought to the makeshift clinic, government and rebel supporters alike, and that brought him to the attention of the Assad regime. Word came to him that he was on the government's wanted list. 'My work with injured people made it look as if I was against the government,' he told us. 'I had a friend in the Syrian army and he said I should take care as they will try to kill me.' He was warned that if the army couldn't track him down, they'd kill his wife and child.

Terkawi decided to leave Syria with his family. He would go to the southern border and make for neighbouring Jordan. As he headed there by car, he asked his well-to-do parents in Damascus to arrange for his wife and child to fly to the Jordanian capital Amman. But he had to be sure he himself wouldn't be stopped at the border. A Syrian official promised that in return for a bribe of €4,000 he would temporarily remove Terkawi's name from a list of people wanted by the government. This would allow him a brief window of time to pass through the frontier without being stopped. But he would have to arrive at the checkpoint within a specified sixty-minute period.

'My heart was bumping at the border post. If they caught me they could kill me.' But the bribe worked: he was not stopped. His passport was stamped when he arrived at the appointed time and he crossed into Jordan without looking back.

That same day his wife and child went to Damascus airport: she was asked why she wanted to go to Jordan. Her Syrian husband worked in the Gulf, she lied, but because the Gulf States wouldn't allow her to visit him, she had to meet him in Jordan. She was waved through and joined Ahmad in Amman. The family made it out just in time. Shortly after the Terkawis left, Assad's secret police called on Ahmad's parents and asked where they were. When the police were told the three were in Jordan, they demanded €1,000 in cash to keep their mouths shut.

Terkawi was soon working in Jordan, treating wounded evacuees from Syria. His resourcefulness won him a job managing a medical centre. He went on to work for the Red Crescent but as a Syrian he never felt welcome in Jordan. So after two years he accepted the invitation of a Syrian opposition group to visit Turkey on a special mission. They needed a good pharmacist to diagnose why Syrian refugee children were suffering severe, even lethal side effects after being vaccinated against measles. By this time Terkawi's wife had another child, born in 2014, and he decided to leave the family in Jordan and return to the country after his assignment. But as he went to board the plane to Turkey in September 2014 he was told that if he left he could not return to Jordan. 'When I asked why, I was told, "You are Syrian."' He phoned his wife and told her to pack and bring the children to the airport.

The next day they were in south-eastern Turkey and Terkawi examined the vaccine that had now killed several children. He immediately spotted that the 'vaccine' was actually an anaesthetic that had mistakenly been put in the same box as the vaccine. A clinic nurse had failed to read the label. The crisis resolved, Terkawi asked his Syrian employers for more work in Turkey. Probably not, was the reply, but they'd ask around and let him know. Hearing nothing and despairing of finding a job in Turkey, Terkawi decided the only hope for his family was to seek asylum in Europe: that meant finding a smuggler to get them there. He moved with the family to Istanbul and met many human traffickers but decided they 'were all liars'. Only when a friend recommended a 'slightly less dishonest' smuggler in Izmir did Terkawi take his family there and enter serious negotiation. Terkawi told us:

> He offered a big boat to get to Greece, two storeys high and sixteen metres long, very safe for the children. It would cost us €5,000 for me and my wife and another €2,500 for the two children. The smuggler said he'd take our money and leave it in his office and return it to us if we didn't get to Greece. I knew I'd never see the money again but I accepted.

On 15 April 2015 the Terkawis went by bus to Bodrun in the south-west corner of Turkey and were met by a driver who took them to a hotel which Terkawi said they found 'very scary' and already full of Syrian and Iraqi refugees. Trying to reassure his wife and his elder boy, who were now very apprehensive, Terkawi agreed to yet another car trip to a house by the sea. He was agonised by his wife's obvious terror at what lay ahead, but told her they had no choice. They stopped on the way to buy life jackets for the children when they were warned there might not be any aboard. It was just as well: when they arrived five men armed with Kalashnikov assault rifles corralled them into a group on the beach and distributed life jackets for adults only.

A small boat lay waiting in the gathering dusk. It was less than half the length promised – around seven metres long and it was to take twenty-five people. 'When I saw the boat,' said Terkawi, 'it was so small and looked so unsafe that I said we wouldn't go, but one of the smugglers prodded me in the back with his gun and said, "You go."' Terkawi's wife and sons cowered under the canopy in the boat's bow as it set off for the Greek island of Pserimos only five miles from the Turkish coast. The sea was lumpy and the boat's rudder was faulty. In order to steer it the boatman shouted at his passengers to shift their weight from one side to the other. While the vessel made its way across the black sea, women prayed and wept and children screamed. As they neared the shore at the end of the one-and-a-half-hour trip, the boat hit a large rock and the smuggler announced that it was too dangerous to go right into the shore. He said everyone would have to swim the last few metres. Most did, but Terkawi refused, begging the man to take the boat right in to the shore as he had two children with him. 'No,' said the smuggler, 'you will swim.'

Terkawi replied, 'In that case you must take us back to Turkey: I left Syria to save my children. I am not going to risk their lives by swimming in the dark.'

Quick as a flash the boatman seized the one-year-old baby and threw him overboard, shouting at Terkawi, 'There, now you will swim.'

Terrified for his child in the pitch-dark water, Terkawi jumped in, leaving his wife and other son in the boat. The baby boy had disappeared under the water. For a few dreadful seconds Terkawi could see no sign of his child. 'But because he was wearing a life jacket, he popped up in front of me. I was worried he'd swallowed salt water but when he moved his legs I knew he was alive. He was crying too. I pushed him towards the beach. Once ashore he didn't cry, he was in such shock.'

The horror wasn't over yet: after throwing in the baby, the boat's driver snatched Terkawi's four-year-old from his wife's arms and threw him in the water too. Without another thought she dived in after her son, and managed to steer him to the beach as well. When they crawled out of the water the child was in tears: 'Why did the man throw me in? I'm so cold.' Everything they had was soaking wet; only the Syrian documents – passports and medical ID cards – that Ahmad had stuffed into the little waterproof pouch he carried in his breast pocket was safe and dry. They had left all their modest belongings in two knapsacks on board as well as food and drink for the children, a bag of nappies and €500 in cash. The boat had vanished into the dark on its way back to Turkey. Its helmsman would keep the belongings his human cargo had left behind and receive a share of the money they had paid for the crossing.

At least, Terkawi and his wife told each other, they had survived a crossing that had killed thousands of others. And they were in Europe, on a tiny Greek island little more than two miles wide. But it was hilly and the going was rough. They had nothing to eat and drink and there was no sign of any help. They spent a cold night in the open: fortunately one of the refugees managed to light a fire. The next day they clambered across the rocky terrain for six hours and at last found a small village. They were welcomed into a house and given a meal, a shower and a night's sleep. The next day Greek police arrived and ferried them across to the larger island of Kalymnos and then to a refugee camp in Athens.

Ahmad Terkawi spent the next four months trying every stratagem to fly his family from Greece to the European Union country of their choice – Sweden. They were lucky on one count: Terkawi's father did business with a Swedish company in Syria and he, the father, and Terkawi's mother had managed to leave Syria and were now living in Sweden. Somehow Terkawi contrived to send his elder child to join his grandparents in Sweden and although he and his wife and baby were still stuck in Athens, they now had every incentive to move heaven and earth to get to Sweden too. It was to cost them a small fortune, which used up most of Terkawi's savings and a loan from his father.

First they had to acquire travel documents that would allow them to fly to Sweden. 'No problem,' promised a smuggler in Athens; 'you look Italian: for €4,000 each I can get you "Italian" passports.' Time and again Terkawi, his wife and child presented the fake documents at the airport without success. Each time the passports were confiscated and each time the smuggler gave them new ones. After their tenth unsuccessful attempt to board aircraft, a sympathetic Greek immigration official took them aside and advised them, 'Tell your smuggler not to use Italian passports next time.' But there was to be no next time at Athens airport. The Terkawis decided that escape route had been barred too often so they embarked instead on a gruelling odyssey through the Balkans starting with a bus trip to the border with Macedonia. It was August 2015.

Macedonia was the smallest country they passed through and the only one to give them no trouble. A train took them to the border with Serbia and there they again ran into the kind of people that made them despair of human nature. Bus drivers offered to take them the 300 miles to Belgrade for €50 each. Twice they boarded buses packed with refugees who'd paid for their tickets, twice the drivers dumped them shamelessly at refugee camps only a mile or two from where they started. They finally made it to Belgrade one morning at 3 a.m. Exhausted, they tried to find a hotel room but no one would take refugees so they slept at the bus station. The following day they took a bus to the Hungarian border, which was closed. They would have to go cross-country with a group of twenty other refugees through fields of maize to a spot where they could cut through the frontier wire fence and sneak into Hungary. At one stage three youths stopped them and asked for money but the refugees felt strong enough to chase them away, shouting they would kill them if they dared to follow. They didn't.

Once through the border fence they started walking, but were soon picked up by Hungarian police who threw them all in jail. They were packed in – ten to a cell – and the food was, said Terkawi, 'uneatable'. At one point, the one-year-old Terkawi baby, who'd just started walking, tottered over to the bars of the cell and gripped them in his tiny hands. 'Get that child away from there,' shouted the guard, 'or I will kill it.' They were told they would be released from jail if they agreed to be fingerprinted. Under European Union law refugees are supposed to seek asylum in the country that takes their fingerprints. Hungary was the last place Terkawi wanted to register for asylum. After four days in the cell, their little boy became ill. 'He was near to death. He was vomiting and had a high fever,' Terkawi told us. So very reluctantly Terkawi agreed to give the Hungarians their fingerprints. Once that was done, the guard threw the family out of the jail shouting, 'Get out of here, you fucking Syrians.' Terkawi then described the family walking along the road towards Budapest trying to hitch a lift. 'No one stopped. Some cars even tried to hit us.'

In the Hungarian capital no train would take them to Austria, so they caught one heading to a station twenty miles from the Austrian border. They walked to the frontier and their Syrian IDs were enough to get them into Austria. As they milled around in a huge crowd waiting for trains to Vienna, a generous Austrian took pity on the couple with the baby boy and offered them a bed for the night. Terkawi, by now suspicious of anyone who offered help, was inclined to say no, but in the end allowed himself to be persuaded. He and his wife could not believe their good luck. The Austrian took them to his elegant three-storey home and his Swedish wife offered them showers and gave them supper. The following morning Ahmad awoke from a ten-hour sleep, and looked at the trees outside the window. 'I thought I had died and gone to paradise.' He was touched when his host asked what Syrians liked for breakfast and promptly drove off to buy yoghurt and cheese.

The next day, 10 September, the Austrian couple bought the Terkawis train tickets to Malmö in Sweden. They were distraught when a Swedish official pointed out that because the family had given their fingerprints in Hungary, they should strictly be applying for asylum there. Fortunately it soon became clear to the Swedes that the Hungarian attitude to Syrian families was so hostile that they agreed the Terkawis could stay temporarily. At the time of writing the family, now reunited with their elder child, is being well cared for in a town not far from Stockholm. They have been granted permanent asylum and the right to look for work. Never one to be idle, Ahmad has set up a group of volunteers who help refugees trying to make the dangerous sea crossing to Europe. They use the internet and telephones to advise on weather and routes. They only assist people who dodge the smugglers by using their own boats. He also coaches a football team in the refugee centre where he lives. He is grateful to the Swedish government for taking his family in, but when we asked where he and his wife would like to spend the rest of their lives, Ahmad Terkawi replied sadly, 'In Homs . . . once peace is restored. But that may not be in our lifetime.'

COMPASSION

COMPASSION IS JUST as central to warfare as courage and cunning. The history of war is not just a list of heroic acts but, equally important, a record of extraordinary selflessness. The battlefield is a place of violence and hatred; but there's altruism, love and sacrifice too. For every war crime there has been an act of kindness equally striking.

As journalists we have witnessed first-hand the death and destruction caused by war. But we have also met exceptional individuals who showed compassion even in the darkest and most dangerous of times. We have broadcast television reports about many of them. We have told the story of an unassuming man in the Netherlands who risked his life by hiding Jews in his basement during the Second World War. We have interviewed a forgiving mother whose family was killed by the IRA. In the midst of her grief, she worked with the assassins in order to bring peace to Northern Ireland. We have talked to a devoted doctor in Tel Aviv who defied his government to give lifesaving care to Palestinian children injured by Israeli shells. At the time of writing we are constantly reminded of the courage of the doctors and nurses who risk death to tend the wounded in the war zones of the Middle East. All these inspirational individuals share similar qualities. They are ordinary, modest and kind. They do not boast about what they do: all say their only instinct is to save human life.

Positive tales of compassion abound in the history of war. Take Richard Kirkland, a young Confederate soldier in the American Civil War. He was at the Battle of Fredericksburg in 1862 when thousands of Union troops were killed and injured. After listening all night to the cries of the wounded, he begged his commanding officer to allow him to carry water to enemy soldiers who were dying of thirst. Firing stopped and both sides cheered as Kirkland bravely marched on to the battlefield with his canteens. He became known as 'The Angel of Marye Heights'. A further model of humanity is German factory owner Oskar Schindler who saved more than 1,000 Jews from the gas chambers during the Second World War. We might never have known how he spent a fortune bribing officials to turn a blind eye to Jewish workers in his factories had it not been for the novel *Schindler's Ark* which was made into the Oscar-winning movie *Schindler's List*.

One of our subjects in this chapter is Nicholas Winton, who also saved Jewish lives. His story might have been lost for ever if scrapbooks containing evidence of his wartime actions had not come to light half a century later. Another compassionate hero who might have been forgotten is John Rabe. He painstakingly recorded in diaries what happened during the ferocious Japanese attack on Nanking, China, in 1937. We tell his story inspired by those diaries that lay for years in an attic, unread. But we begin this chapter with a woman who made an extraordinary contribution to caring for the wounded in two World Wars. Both our grandfathers fought in the First World War, and passed on terrifying tales of gas attacks and rat-infested trenches. We know from them what it was like to fight. We know much less about the many brave women who, in their different ways, gave so much to the war effort. Which is why we find Mary Borden's story so fascinating.

29

Mary Borden

Nurse and Novelist, 1914–18 and 1939–45

ONE OF THE few ways that a woman could get to the front in the First World War was as a nurse. Some wrote diaries and memoirs chronicling the horrors of that war but no one captures its stark brutality like Mary Borden. In *The Forbidden Zone*, a book she wrote while nursing at the front, she paints a haunting picture and doesn't spare the gory details: 'There are heads and knees and mangled testicles. There are chests with holes as big as your fist, and pulpy thighs, shapeless; and stumps where legs once were fastened. There are eyes – eyes of sick dogs, sick cats, blind eyes, eyes of delirium; and mouths that cannot articulate; and parts of faces – the nose gone, or the jaw.'

It's not surprising that military censors banned publication of her horrifyingly explicit account until after the war. Her vivid writing is not the only reason Mary Borden appeals to us. She also set up and financed field hospitals in both World Wars. And just how she ended up in the middle of those conflicts is a tale stranger than fiction.

Little in her early life suggested that she would become a leading figure in caring for the hideous casualties of war. Born in Chicago in 1886, Borden inherited a fortune when her millionaire father died. She was just twenty when she became an heiress but she already had a strong social conscience and a zest for adventure. On a round-the-world trip with family friends, she fell in love with Douglas Turner, a young Scottish missionary she met in India. The novelist E. M. Forster described him as a 'charming man, most sympathetic and full of fun'. Turner proposed to Borden and they married in 1908, just five months after they met. The couple settled in London in 1913 and Borden wrote her first novel, *The Mistress of Kingdoms*. Its heroine was headstrong and vibrant just like its creator.

Borden threw herself into London's social and political life. Setting herself up as a literary hostess, she held star-studded parties for writers such as Forster and George Bernard Shaw. She joined the suffragettes and ended up in jail for five days after throwing stones and breaking a window during a Votes for Women demonstration outside Parliament. A small woman, just five foot two, she made sure she was seen and heard. With her dark, penetrating eyes and lively, expressive face she stood out even in the most cerebral of gatherings. Friends described her as 'exquisitely elegant and chic, intelligent, quick and witty, knowledgeable about politics and many international problems'.

When Britain went to war in 1914, Borden quickly realised that, instead of throwing parties or rocks at demonstrations, she could use her money and her gift for organisation to save lives. She volunteered with the London Committee of the French Red Cross. Her account of a meeting with the charity's president, La Vicomtesse de la Panouse, is almost comic.

> The Vicomtesse asked:
> Had I done any nursing?
> No.
> Did I speak French?
> A little.
> There was a typhoid epidemic in Dunkirk. Was I willing to nurse typhoid?
> After a moment's hesitation, once more, yes.

Would I go, taking two hospital nurses with me at my own expense? Yes.

At the end of January 1915, in spite of having just given birth to her third daughter, Borden headed to Dunkirk. The German army now occupied most of Belgium but had not managed to overrun all of northern France. When she arrived in the port she was directed to the run-down casino of Malo-les-Bains which had been turned into a makeshift typhoid hospital. 250 patients lay helpless under the great chandeliers of the roulette rooms.

'The rows of dingy beds were reflected to infinity in the vast gilded mirrors,' writes the heiress-turned-nurse. 'The stench in the rooms made one vomit. I would run every so often behind my screen to be sick then hurry back to that dim purgatory of gaunt heads, imploring eyes and clutching hands.'

Borden soon learned how to give injections and dress wounds but she and the two British nurses she had brought with her struggled with French army regulations which, among other things, did not allow night nursing. When Borden complained to a visiting inspector he suggested that she start her own hospital. Impulsively, she wrote to General Joseph Joffre, French commander-in-chief, offering to equip and run a field hospital for 100 patients. To her surprise and delight he said yes.

And so in July 1915 Borden's l'Hôpital Chirurgical Mobile No 1 opened near the Flanders village of Roesbrugge, not far behind the front line of Ypres. The French government supplied doctors, medical supplies, food and transport. Borden paid for the building and nurses. To help meet running costs, she wrote to American newspapers asking for contributions and begged family and friends for support.

Borden was in her element. She writes of her sense of 'great power, exhilaration and excitement'. She was immensely proud of her hospital, which she describes as 'throbbing and humming at night like a dynamo. The operating rooms were ablaze; twelve surgical *équipes* were at work, boilers steamed and whistled; nurses hurried in and out of the sterilising rooms, carrying big shining metal boxes and enamelled trays; feet were running, slower feet shuffling. The hospital was going full steam ahead.'

Her hospital was a great success. In just six months it treated 800 seriously wounded men. With a mortality rate of just 5 per cent – the lowest on the Western Front – the hospital became so renowned that officers demanded to be treated there. Borden's writing reflects how upsetting it was to help patients recover only to send them back into battle.

> This is the place where he is to be mended. We lift him on to a table. We peel off his clothes. We cut off his shirt with large scissors. We experiment with his bones, his muscles, his sinews, his blood. We dig into the yawning mouths of his wounds. To the shame of the havoc of his limbs we add the insult of our curiosity and the curse of our purpose, the purpose to remake him. It is only ten kilometres up the road, the place where they go to be torn again and mangled.

After nursing for hours on end, Borden went to sleep to the sound of guns booming in the distance. She dealt with rain that came through holes in the roof by covering her pillow with an umbrella. When she went on leave to visit her family, now living in Paris, she complained that the silence kept her awake.

Borden's mission to save lives was to be put to the test in the summer of 1916 when the French and British launched a massive and largely futile offensive on the River Somme. On 1 July, the first day of the infamous battle, 19,000 British soldiers died and 41,000 were wounded. It was the largest one-day loss of life in the history of the British army. With the French suffering casualties too, Borden offered to set up a second hospital. She was refused permission on the grounds that she was not French and instantly fired off another letter to General Joffre. She might be American, she told him, but she felt French at heart. Her emotional appeal worked and in October 1916 she arrived at a hastily built hospital in Bray-sur-Somme, just three miles from the front. Her job was to finance and run what would become the French army's most important military hospital.

Borden's first impression of her new project was not promising. With its rows and rows of small, badly constructed buildings she said it reminded her of 'an American lumber town. A city of huts, and the guns beyond the hill sound like the waves of the sea, pounding – pounding – and the sky is a-whirr with aeroplanes, and sometimes we are bombarded and all the time troops and more troops stream past.' Borden called the hospital 'a second battlefield'. When the fighting was particularly fierce stretchers covered every available space. It was her job to meet ambulances and decide who needed care most urgently:

> If a man were slipping quickly, being sucked down rapidly, I sent runners to the operating rooms. It was my business to know which of the wounded could wait and which could not. I had to decide for myself. There was no one to tell me. If I made any mistakes, some would die on their stretchers, on the floor, under my eyes who need not have died. I didn't worry. I didn't think. I was too busy, too absorbed in what I was doing.

Later, looking back, she wonders how she coped: 'I think that woman, myself, must have been in a trance, or under some horrid spell. Her feet are lumps of fire, her face is clammy, her apron is splashed with blood; but she moves ceaselessly about with bright burning eyes and handles the dreadful wreckage of men as if in a dream.'

Somehow in the midst of all this carnage, Mary Borden met the man of her dreams. Her marriage had not been a happy one. While she greatly respected her kind and supportive husband, she was not physically attracted to him. One day a handsome twenty-eight-year-old officer whose job was to liaise between British and French forces turned up at the hospital looking for missing comrades. Captain Edward Louis Spears remembered being astonished to find a woman so near the front, especially such a small one in a bloodstained apron. He soon fell for the 'bright, brave lady' and they started a passionate affair. Borden called it 'love at first sight'. They wrote to each other every day and met whenever possible.

By 1917 Borden was back at her first hospital near Ypres which was now coming under regular attack. With typical sangfroid she told an interviewer from the *New York Herald*: 'Being under fire really means nothing, for one soon learns how to avoid being hit.' But she did admit it was: 'really terrifying, when you hear houses being crushed all about you, and you never know but at that moment you, too, may be crushed'.

The French who had been so reluctant to let Borden set up her hospitals now rewarded her for her generosity, courage and for the huge contribution her hospitals made. She was given not just one but two of the highest French honours: the Croix de Guerre and the Légion d'Honneur.

Her work was far from over. The late summer and autumn of 1917 were unusually wet, which proved disastrous for British Field Marshal Douglas Haig's new offensive, the Third Battle of Ypres. Men, animals, guns and tanks became bogged down by mud. It was trench warfare at its very worst. The death and devastation is palpable in lines from Borden's poem, 'This is the Song of the Mud'.

> This is the hymn of the mud – the obscene, the filthy, the putrid,
> The vast liquid grave of our armies. It has drowned our men.
> Its monstrous distended belly reeks with the undigested dead.
> Our men have gone into it, sinking slowly, and struggling and slowly disappearing.

Three years after she and her team had started their work of saving thousands of lives at the front, Borden became a patient herself. At the end of 1917 she was diagnosed with a lung infection probably caused by caring for soldiers with gangrene. Her month-long convalescence was interrupted by a divorce hearing. Her husband had discovered her affair with Spears and reluctantly agreed to end the marriage. In February 1918 Borden miscarried Spears's child. Two months later the lovers were married in Paris.

When the war ended Borden moved as seamlessly into peacemaking as she had into saving lives. She opened her Paris home to all the main players at the peace conference that led to the Treaty of Versailles. She took great delight in bringing together politicians such as David Lloyd George and Winston Churchill with poet and artist Jean Cocteau, the controversial hero T. E. Lawrence and the young economist John Maynard Keynes. Churchill later became godfather to her son Michael.

In the years before the Second World War, Borden added to her accomplishments by becoming a bestselling novelist. A critic in the *Telegraph* called her 'one of the most vivid and engrossing figures that have dazzled the book world for many years past'. Back in London she took her reputation as a generous and entertaining hostess to new heights, making friends with writers like Noël Coward. He described Borden as a 'small attractive woman with deep sleepy eyes and a rather nervous smile'.

In addition to her parties and novels, the indefatigable Borden wrote plays, short stories, newspaper articles and campaigned for women's rights and penal reform. But when Britain declared war on Germany on 3 September 1939, her dynamism and compassion prompted her to spring into action again. Aged fifty-three, she joined a London ambulance unit. With the full support of her husband, Edward, who was soon to be appointed Prime Minister Winston Churchill's special representative to France, Borden once again became immersed in wartime life.

A few weeks into the war she wrote: 'Only slept 3 hours last night. I lose track of the days and time. It seems years since Germany invaded Poland. As if with all one's habits violently broken and not only the fabric of one's little life – but of one's world, space, time, all the rest of it – was shattered.'

When an old friend, Frances, Lady Hadfield, the wife of a British steel tycoon and an ardent Francophile, offered to give Borden a large sum of money to help the French war effort, she lost no time setting up another field hospital. She named it Hadfield-Spears Ambulance and it was assigned to the French Fourth Army in the north-east of France bordering Germany. Borden quickly assembled the necessary tents, surgical equipment and supplies, X-ray machines and portable kitchen. The unit was up and running by February 1940.

Three months later, with the Germans' rapid advance through Belgium into France, the Fourth Army and its hospital were ordered to evacuate. Borden and her medical team headed south. The French military collapse was so swift that plans to set up her field hospital en route had to be abandoned. Her unit crossed France without a map, short of petrol, navigating roads full of fleeing refugees. Finally, in order to save her British nurses, she reluctantly abandoned her medical equipment and her French orderlies and doctors. Before running for her life, she shook hands with each of them, thanking them for all they had done.

Packing her twenty-five British nurses into six vehicles, she headed for Bordeaux but was stopped at a roadblock just outside the city. She was told the road was closed and she could go no further. 'I must,' she insisted. 'These ladies are all British. We are obliged to leave France.' Once again her forceful personality prevailed and the little convoy was waved through. It drew up outside the British consulate and Borden pushed past a waiting crowd. 'I am Mrs Spears,' she announced to the last British official there, a military attaché who was busily burning the consulate's papers. 'We have been so worried about you,' he replied with relief. After two nights waiting nervously to be evacuated, with the help of naval Captain Ian Fleming, later famous as the creator of James Bond, she and her nurses were smuggled aboard two fishing boats. They were transferred to the last passenger ship leaving France. When the exhausted group arrived back in England on 26 June 1940, they discovered newspapers had reported them 'lost without trace'.

A week before she left France, Borden's husband had an equally dramatic dash to Britain with none other than General Charles de Gaulle – later commander of the Free French. They were united in fury at how quickly the French had surrendered. De Gaulle was convinced he was about to be arrested by the pro-German Vichy government and Spears offered to help him escape from Bordeaux. The general accompanied Spears to the airport, supposedly to bid him farewell, but as the plane moved off, de Gaulle grabbed Spears's arm and leapt aboard. The two men worked closely together in the months that followed but Borden, who saw a lot of de Gaulle, made no secret of the fact that she didn't trust the self-appointed leader of the Free French. In her memoirs she writes: 'He could make himself very agreeable if he felt so inclined. But he was often biting, scathing in his criticisms of England and the English, just as much or more so of France. His long lips would grimace as if he were drinking gall and wormwood when he talked of France.'

In spite of her feelings, she was thrilled when de Gaulle accepted her offer of a field hospital to accompany his troops to the Middle

East. With the help of a generous contribution from the British War Relief Society in New York, Borden put together her most ambitious medical unit yet – a 100-bed hospital that could be packed up at almost a moment's notice and moved to where it was most needed. When the reconstituted Hadfield-Spears Ambulance was shipped from Cardiff in February 1941, the convoy carrying its equipment was two and a half miles long.

Borden joined the unit in Palestine just in time to supervise its move to the Syrian border. The British and de Gaulle's Free French were determined to crush Germany's allies, the Vichy French, who controlled Syria and Lebanon. No sooner had the hospital tents gone up than injured Free French soldiers started being brought in. They had been wounded in a battle with the Vichy army. Borden was appalled that they were suffering at the hands of their own countrymen. 'I feel again, not the physical suffering of the men's mangled bodies but the festering pain of their minds. And as they tossed and writhed in their beds, as they raved in delirium, as they died, I know that one thought tormented many, namely, that this had been done to them by their own people.'

Over the next few months the Hadfield-Spears unit followed the fighting from Damascus to Beirut. Hospital tents were set up under trees, in a convent, by the side of desert roads. Between finding locations, running the unit, dealing with staff conflicts and nursing, Borden was too frantically busy to write.

After only two months the Allies were victorious and Borden's husband became British Head of Mission to Syria and Lebanon. Her life became one of stark contrasts. One week she was in what she later described as 'suffocating, fly-infected schoolrooms smelling of blood and gangrene' and the next she was in Beirut dining with the President of the American University.

In early 1942 Borden temporarily abandoned her elegant dinner parties and joined the North African desert war where Britain and its allies were fighting the Germans and Italians. Her medical unit moved with the First Brigade of the Free French, who were supporting the newly formed British Eighth Army, as the front swung to and fro in Egypt and Libya. Somehow Borden and her English nurses, who doubled as drivers, made it through sandstorms, careering along obscure tracks in the sand. They even went across a minefield as they followed the French brigade in the fight against the resourceful German General Erwin Rommel. Borden agreed with an officer who said it was crazy to have ladies in the desert: 'If he meant our driving across the desert through a sandstorm alone,' she later wrote, 'he was right.'

Her desert life was interspersed with trips back to Beirut where her husband, now Sir Edward, had been elevated to the post of ambassador. Acting as his hostess, she impressed guests like Hermione, the intrepid Countess of Ranfurly, who was mesmerised by Borden's war stories and wrote: 'Her voice is high and she speaks very slowly but one never gets impatient because she is so interesting. Lady Spears has a lot of courage – that sticks out a mile.'

Borden's Hadfield-Spears Ambulance served the Allies across North Africa and into Europe through Italy. By the summer of 1944 it was back in France. But a year later after Germany's defeat its gallant war record got short shrift at a victory parade in Paris. General de Gaulle had fallen out with Spears, his former close friend. And when he saw the word 'Spears' on the sides of Borden's ambulances and British flags on her cars and trucks, he was so angry that he ordered the unit closed down within forty-eight hours and all British staff members sent home. Borden never forgave him: 'Neither then, nor any time during the war, did he say anything that could be construed as a tribute to the work we had done.'

When peace came, Borden, who had two more decades to live, returned to her busy civilian life, writing and socialising with friends and family. She even helped her nephew-in-law Adlai Stevenson with his political speeches when he ran for US President in the 1950s. But to the end of her long life she was very clear about where she had found her true vocation. It was, she said, in the two World Wars looking after wounded and sick soldiers that she felt most alive.

30

John Rabe

The Living Buddha of Nanking, 1937–8

THIS IS THE story of one man's humanity in the midst of one of history's worst massacres. For six weeks from mid-December 1937 the Chinese capital of Nanking was a bloody killing ground. Young Japanese soldiers ran amok. They looted, burned, raped and murdered, often encouraged by officers who believed such behaviour would turn them into battle-hardened warriors. Photographs show the marauding soldiers smiling broadly as they wave heads they've just cut off and parading proudly among piles of dead bodies. Estimates of the number of civilians killed by the Japanese range from 260,000 to 350,000. As many as 80,000 women and children are thought to have been raped. This orgy of violence was one of the most brutal and disturbing episodes of modern warfare. But the death toll would have been even higher had it not been for the actions of a group of brave westerners led by a German businessman named John Rabe.

It was Imperial Japan's craving for more territory and resources that led it to invade China in 1937. Decades of hostility towards the Chinese whom the Japanese both resented and despised exploded into total war. And unassuming, good-natured John Rabe found himself in the middle of it. Rabe, a balding, bespectacled German, had lived in China for more than thirty years and was a popular figure in the bustling city of Nanking. He worked as chief salesman for Siemens engineering which supplied turbines for the growing city's electrical plant, equipment for its hospitals and its new telephone exchange.

Rabe was more than an expatriate businessman. He was an active member of the community. He built a German school in the grounds of his small but comfortable home. He also headed the local Nazi Party. Hitler's regime in Germany was only four years old and in those early days Rabe, like many Germans (and indeed we have to admit other westerners too), saw the Führer as a progressive man of peace. Rabe had a deep affection for the people of China and their culture and was appalled at the savagery of the Japanese onslaught. He was also a keen observer, and kept a diary most of his life. Between September 1937 and April 1938 he filled seven notebooks. They provide a horrifying, day-by-day account of what became known as the Rape of Nanking.

In August 1937, before the Japanese reached Nanking, Rabe took a holiday from the suffocating heat of the city and went to a Chinese coastal resort with his wife, Dora. Most people expected that Chiang Kai-shek's Nationalist forces would hold off the Japanese for months if not years. But when Rabe learned that Japanese planes were dropping bombs on Nanking, he left his wife in the safety of the resort and rushed back to make sure his business and its employees were safe. On 21 September, as most other foreigners, terrified of being attacked by the Japanese, streamed out of the city, Rabe writes in his diary: 'Can I, may I cut and run? I don't think so. Anyone who has ever sat in a dugout and held a trembling Chinese child in each hand through the long hours of an air raid can understand what I feel.'

By the time he wrote these words, the Japanese war machine appeared unstoppable. What began with its occupation of Manchuria in northern China in 1931 had developed into an all-out campaign to conquer the whole country. After the fall of Shanghai in November 1937, it appeared to be only a matter of time before the capital city of Nanking would fall. In his diary Rabe admits: 'I haven't the least desire to put my life at risk.' But he felt he could not abandon his employees or the hundreds of thousands of Chinese who could not afford to flee. 'Aren't they in danger of being slaughtered in great numbers? Shouldn't one make an attempt to help them? Save a few at least?'

Determined to make a difference, Rabe came up with a plan. He knew that the world's first safety zone for civilians had been set up in Shanghai by Robert Jacquinot de Besange, a compassionate French priest, and that it had saved thousands of lives when that city fell to the Japanese. On 19 November Rabe attended a meeting of an international committee of sixteen people, mainly American missionaries and doctors. Its aim, he said, was 'to try to create a refugee camp, or better, a neutral zone inside or outside the city, where non-combatants can take refuge in case the city comes under fire'.

When the committee asked Rabe to be its chairman, he reluctantly accepted. He was worried that he might not prove a worthy leader but those who knew him had no doubts. An American doctor on the committee wrote that although Rabe was well up in Nazi circles, 'discovering what a splendid man he is and what a tremendous heart he has, it is hard to reconcile his personality with his adulation of "Der Führer".' Rabe's German nationality and his Nazi links made him a valuable asset, as Nazi Germany and Imperial Japan were growing closer. They had jointly signed a pact against Soviet Russia a year earlier and the two countries viewed each other as useful allies.

As the Japanese army moved ever closer to Nanking, Rabe threw himself into organising a two-and-a-half-square-mile section of the city which he hoped would remain safe from attack. The borders of the area, marked by white flags and sheets painted with circles containing red crosses, were fiercely guarded by members of the committee. Most experienced Chinese soldiers had been ordered to retreat from Nanking, leaving young recruits to fight the Japanese. Many of them tried to push their way into Rabe's safety zone, which would have compromised its neutrality. As well as trying to expel these men, Rabe had to struggle to find accommodation for 250,000 civilians who crowded into the zone (around half the population of

Nanking). The tireless German oversaw food distribution, sanitation, health care, policing and finances. He played the diplomat too, sending a telegram to Hitler urging him to encourage the Japanese government to respect the safety zone. Hitler didn't reply but Rabe was convinced the Führer got his message because soon afterwards Japanese planes stopped attacking civilian targets. This impressed his American colleagues no end, Rabe wryly notes. He also sent messages to the Japanese government begging them to spare the safety zone from attack. Each request was turned down.

Rabe's diary entries in the days leading up to the Japanese army's arrival reflect increased urgency: '6 December 1937: If those of us in Nanking want to be truly prepared, we must have rice flour, salt and fuel, medicine, cooking utensils.'

Before the Mayor of Nanking, along with other Chinese officials, fled the city, he presented Rabe and his committee with 2,000 tons of rice and 10,000 bags of flour but they were piled up in warehouses outside the city. Trucks that were supposed to move the food into the safety zone were commandeered by the hard-pressed Chinese to evacuate their troops and remove treasures from museums. Undeterred, Rabe and his fellow committee members drove their own cars to the warehouses and crammed them with food. They only managed to move a fraction of the huge stockpile but every grain of rice made a difference.

As more and more people sought safety in the zone, Rabe helped arrange for schools and universities to provide shelter. He also offered his own house and garden to the arriving masses. On 12 December, the day before the Japanese entered Nanking, he writes: 'women and children plead to be let in . . . And I can't listen to their wailing any longer, so I open both gates and let everyone in who wants in.' Rabe made space in every corner of his house including the coal bin and the servants' lavatory. He found room in garden sheds and on his lawn. In spite of cold, winter weather, the refugees were happy to sleep out in the open. The most popular spot was under a huge Nazi flag that Rabe had stretched out in case of air raids.

Rabe had no idea whether the Japanese would respect his safety zone but it had worked in Shanghai so he remained optimistic. Besides, Chinese friends told him the Japanese would restore order quickly once they took over Nanking. His last thought on the night before the city fell was 'Thank God that the worst has been overcome.' He could not have been more mistaken.

From day one of the Japanese occupation of Nanking, there was bloody mayhem. Rabe writes: 'We come across corpses every 100 to 200 yards. The bodies of civilians that I examined had bullet holes in their backs. These people had presumably been fleeing and were shot from behind.' Thousands of retreating Chinese soldiers rushed into the safety zone and Rabe persuaded many of them to lay down their weapons and surrender. He wrote a letter to the Japanese military commander demanding that the defeated soldiers be treated humanely, according to international rules of war. He was shattered when Japanese soldiers started rounding up the men he had helped disarm. He watched helplessly as 500 Chinese soldiers were marched off. 'We assume they were shot since we later heard several salvoes of machine-gun fire. These events have left us frozen with horror.'

Two days later, Rabe and two American colleagues tried valiantly to negotiate the safe release of 1,300 Chinese soldiers: 'But to no avail. They are surrounded by about 100 Japanese soldiers and still tied up are led off to be shot . . . I'm in a truly wretched mood. It's hard to see people driven off like animals.' The Japanese examined all captives for signs of military activity. If they had calluses on their hands that might have been caused by regular use of guns, if their heads revealed marks of military caps, if their backs showed signs of carrying a knapsack, they were immediately shot or beheaded. Thousands of civilians who did manual work that caused hand calluses or had backs marked by carrying heavy loads were also brutally shot, bayoneted or burned to death. Japanese soldiers used Chinese prisoners for bayonet practice. They staged decapitation contests and competitions to see who could kill 100 people most quickly with a sword.

Desperate to stop the wholesale killings, Rabe tried to capitalise on Germany's growing friendship with Japan and wrote to staff at the Japanese Embassy in Nanking. He listed each atrocity and called for official action to stop the slaughter. Each time the reply came back: 'We shall inform the military authorities.'

As well as hectoring Japanese officials and overseeing the

management of the safety zone, Rabe also acted as chief protector to the 650 people who were camped out in his house and garden. On day three of the occupation, 16 December, he writes: 'All the women and children, their eyes big with terror are sitting on the grass in the garden, pressed closely together, in part to keep warm, in part to give each other courage. Their one hope is that I the "foreign devil", will drive these evil spirits away.' With Japanese soldiers roaming the streets looking for women to rape, 'the foreign devil' gave his female guests whistles to blow. He slept in his boots ready to leap out of bed when the whistles sounded. His guests adored Rabe. He didn't just protect them. He was unfailingly generous. When women had babies in his garden he gave them a gift of money. They duly named their children John and Dora (after Rabe's wife). 'Great fun,' he writes. On day four of the occupation, he sent four Japanese soldiers packing. 'They don't want to tangle with a German. Usually all I have to do is shout "Deutsch" and "Hitler" and they turn polite whereas the Americans have real trouble getting their way.'

Two days later he describes soldiers clambering over his garden wall and trying to open the gates to let their comrades in: 'When I arrive and shine my flashlight in the face of one of the bandits, he reaches for his pistol, but his hand drops quickly enough when I yell at him and hold my swastika armband under his nose. Then on my orders, all six have to scramble back over the wall . . . at least 20 attempts to get in so far.'

Eleven days after the fall of Nanking, Rabe visited a hospital in the safety zone and saw for himself the results of the killing frenzy. He met a sampan owner on the verge of death who'd been shot and then burned. He went to the morgue and saw recently delivered bodies:

> among them a civilian with his eyes burned out and his head totally burned, who had likewise had gasoline poured over him by Japanese soldiers. The body of a little boy, maybe seven years old, had four bayonet wounds in it, one in the belly about as long as your finger. He died two days after being admitted to the hospital without ever once uttering a sound of pain.

Rabe was tormented by such suffering but he was determined to see for himself so he could tell the world what went on: 'A man cannot be silent about this kind of cruelty.'

As more and more brutality came to light, Rabe redoubled his efforts with the Japanese Embassy. The carefully compiled lists of atrocities he handed to embassy staff disappeared into unknown hands but, politely and firmly, Rabe persisted. On 16 January 1938 he and some of his committee colleagues were invited to dinner at the embassy. Rabe made a speech thanking the Japanese for patiently hearing 'our requests and complaints'. He later wondered what the Americans thought of his thanks. 'I spoke a little against my own conscience, but I thought it useful for our cause and followed the Jesuit principle: "The ends justify the means".' The practical Mr Rabe knew that without the help of Japanese officials, his attempts to end the atrocities would get nowhere.

It wasn't just killings and rapes that concerned him. Every hour of every day, Chinese men, women and children were treated abominably in a myriad of ways. He writes about 'a Chinese man who worked all day for the Japanese and was paid in rice. His wife prepared it for him and their hungry children and as they ate it a Japanese soldier passing by plays a little joke and urinates in the half-full bowl of rice and laughs as he goes his merry way.'

On 31 January, Chinese New Year, Rabe experienced a moment of rare joy. His refugees presented him with an enormous banner covered with the words:

> You are the Living Buddha for a hundred thousand people.
> You have the heart of a Buddha
> And share his bold spirit.
> You have saved thousands of poor people
> From danger and want.
> May the favour of Heaven be granted to you,
> May good luck follow you,
> May God's blessing rest upon you!
> From The Refugees of Your Camp.

'If these were not such perilous times, I could almost laugh at this touching dedication. But I didn't dare take any real delight in this gift . . . You grow weary in this constant battle against a demoralised Japanese soldiery!'

As Japanese soldiers continued their bloody rampage, Rabe and his fellow committee members, at huge personal risk, travelled around the safety zone, trying to restore order. His car was often stopped by Chinese desperate for help. 'The Japanese had pistols and bayonets and I . . . had only party symbols and my swastika armband.' But the courageous German never hesitated to rush to the rescue. He saved many women from being raped: 'The mother of a young attractive girl called out to me, and throwing herself on her knees crying, said I should help her. Upon entering [the house] I saw a Japanese soldier lying completely naked on a young girl, who was crying hysterically. I yelled at this swine, "Happy New Year" and he fled from there, naked and with his pants in his hand.'

Rabe developed the knack of using humour as an antidote to despair. Shouting 'Happy New Year' to scare off a rapist made him laugh but, as days passed and more and more women were abused, he lost heart. 'You can't breathe for sheer revulsion when you keep finding the bodies of women with bamboo poles thrust up their vaginas. Even old women over 70 are constantly being raped.'

Many women were killed after being raped. Their bodies joined growing piles of corpses littering the streets. On one occasion Rabe urged Japanese officials to do something about 1,000 corpses lying unburied. 'These corpses have been partially eaten by dogs. At the same time, however, dog meat is being sold by the Chinese in the streets.' Countless bodies were also thrown into Nanking's many ponds. On 7 February Rabe watched 124 bodies being recovered from two ponds, half of them civilians. The victims' hands were tied. They'd been mowed down by machine guns, covered in gasoline and set on fire. 'But when the burning took too long, the half-burned bodies were simply tossed into the ponds. Another pond nearby is said to contain 23 corpses, just as all the ponds in Nanking have been similarly contaminated.'

After six weeks of chaos, the Japanese authorities slowly started taking action. Ponds and streets were cleaned up, rapes and killings decreased and order was gradually restored. Worn out by all he'd done and seen, Rabe decided he could safely leave Nanking. In a formal declaration, his fellow committee members praised his work: 'Mr Rabe's leadership in difficult tasks of benevolence has been fearless and gentle. It will long be remembered by the entire population of Nanking.'

He travelled to Shanghai on 23 February. Reunited with his wife, he returned to Berlin where he lectured about the atrocities in Nanking. Here's a typical excerpt from one of his speeches:

> Groups of 3 to 20 marauding soldiers would begin by travelling through the city and robbing whatever there was to steal. They would continue by raping the women and girls and killing anything and anyone that offered any resistance, attempted to run away from them or simply happened to be in the wrong place at the wrong time. There were girls under the age of 8 and women over the age of 70 who were raped and then, in the most brutal way possible, knocked down and beaten up. We found corpses of women on beer glasses and others who had been lanced by bamboo shoots.

He illustrated his talks with a film taken by an American missionary in Nanking. It showed bayonet and rape victims, dead Chinese soldiers and civilians. He wrote to Hitler asking for an audience. He was sure the Führer would be so appalled by what had happened in Nanking that he'd complain to the Japanese government. When he received no reply, Rabe sent the Nazi leader a copy of his lecture along with the film. Within days he was visited by Gestapo officers who questioned him for several hours. The German government was now openly aggressive towards neighbouring countries in Europe and valued its growing alliance with Japan. The Gestapo warned Rabe never to talk about Nanking again in public.

John Rabe and his family survived the Second World War in Berlin but he was forced to sell off his beloved collection of Chinese treasures to pay for food and lodging. After the war news reached Nanking that the man who had been worshipped in China as 'the Living Buddha' was living in poverty; Chinese officials promptly arranged for food parcels to be sent to him. He was offered an apartment and pension in China on the condition that he gave evidence at the Japanese War Crimes Tribunal. He turned down the offer because: 'I didn't want to see any Japanese hang, although they deserved it . . . There must be some atonement, some just punishment but in my view the judgement should be spoken only by their own nation.'

Two Japanese generals were found guilty of Nanking war crimes and sentenced to death in 1948. But Prince Yasuhiko Asaka, a member of the Japanese royal family who commanded the attack on Nanking, was never called to account. Under the peace deal that Japan signed at the end of the Second World War members of the imperial family were granted immunity from prosecution.

Rabe died of a stroke in 1950 and his diaries were left to his son Otto. They sat unread, gathering dust in the attic of his Black Forest home. Then in 1991 Rabe's heroism was mentioned in a book published in Germany. The author, Erwin Wickert, assumed the diaries had been confiscated by the Gestapo. But when Rabe's granddaughter Ursula contacted him and told him the diaries still existed, he asked to publish them. She agreed, explaining the family had thought of destroying the manuscripts after the war because after all the horror they had been through, they didn't want to read about more of it.

In 1997, nearly five decades after he died, Rabe's gravestone was moved to Nanking (renamed Nanjing). Today it sits in a place of honour there in the Massacre Memorial Hall. His home in the Chinese city has become a museum and centre for peace. It is a lasting tribute to the man who lived by words he wrote during the Rape of Nanking: 'If you can do some good, why hesitate?'

There is no exact figure for the number of people John Rabe helped to survive the hellhole of Nanking. Estimates put it at up to a quarter of a million.

31

Nicholas Winton

Children's Saviour, 1939

ON 30 JUNE 1939 eleven-year-old Joe Schlesinger and his nine-year-old brother Ernie kissed their mother goodbye on the platform at Bratislava station in Czechoslovakia. They had every reason to believe the parting was only temporary. 'We are sending you away until things blow over,' she told them reassuringly.

In spite of Hitler's promise at Munich to leave most of Czechoslovakia independent, Germany had invaded the country three months earlier. Worried about the future, Joe's Jewish parents were sending their sons to stay with uncles who'd moved to England. Their father was accompanying them as far as Lobositz, a town just inside Nazi Germany where they would join a train from Prague full of children heading to London. Joe remembers his mother waving them off with a smile on her face 'making the best of it, not wanting to scare us kids'. The train they were joining in Lobositz was hours late. In Hitler's Germany Jews were not allowed to enter the waiting room, so Joe, Ernie and their father spent the night in the smelly station lavatory. The next morning, as he put them on the train to

London, his father's parting words to Joe were: 'Be a good boy and take care of Ernie.' The children would never see him or their mother again. Both parents died in the gas chambers at Auschwitz.

There were 237 other children who said goodbye to their parents and left on that train. Sisters Vera and Eva Diamant, aged nine and fourteen, had boarded at the start of its journey. Vera recalls arriving at Prague's Woodrow Wilson station 'in the middle of the night with the platform full of anxious parents and German soldiers'. Before the train pulled out Vera's mother and father smiled bravely but her last memory was of them 'crying – though they still tried to smile as they waved us goodbye'. It was the last time the girls saw their parents. They too died at the hands of the Nazis.

Brothers Hans and Gert Feige, seven and ten, joined the train in Hannover. Their Jewish father was German, their mother Czech and the family lived in Dresden. One of Hans's earliest memories was of other German children in a playground spitting at him and calling him a dirty Jew. He still bears a scar on his forehead – the legacy of a family doctor who examined his head after an accident and said, 'The cut needs stitches but I don't stitch Jews.' The family fled to Czechoslovakia in 1939. After the Germans invaded that country weeks later, their parents sent Hans and Gert to a Jewish school in Hannover which acted as a clearing station for children going on to other countries. Hans remembers his mother being very composed as she waved goodbye. She said she would see him and his brother soon. But at the last minute she handed him her wristwatch 'for you to remember us by'. Both parents died at Auschwitz.

So how did these children escape the dreadful fate the Nazis inflicted on their parents? And who helped them to safety?

We have known Joe Schlesinger for many years. It was when Ann worked with him at the Canadian Broadcasting Corporation that he told her he had been one of the children who'd made it out of Czechoslovakia. Between March and August 1939 there were eight separate mass evacuations by train known as Kinder – or Children – transport. The trains went from Prague to the English Channel where the children could be taken on to London. The Czech Kindertransport trains carried 669 mainly Jewish children to safety. For nearly half a century none of those who made it to England, including Joe, had any idea of the identity of the extraordinary man who organised their rescue.

In 1988 Nicholas Winton and his wife Grete were clearing out the attic of their family home in Maidenhead, near London. In a dusty old suitcase Grete came across a scrapbook. Page after page contained neatly pasted photos of children like Joe Schlesinger, Vera Diamant and Hans Feige. There were letters, documents and newspaper articles. When she asked her husband what it was all about, he told her a story he'd hardly mentioned in their forty years of marriage. These were records of something that had happened before they'd even met, records of the Kindertransport programme that he had set up before the war. She was thrilled to be reminded of something of which she'd been only dimly aware and persuaded her husband to allow her to make the story public. Winton himself had always dismissed as 'unremarkable' what others recognised as a truly striking act of compassion. Grete's fascination with the scrapbook led to a newspaper article and the story soon became a media sensation that brought Winton widespread public acclaim.

Nicholas Winton, known to everyone as Nicky, came from a wealthy Jewish family that had immigrated to England from Germany in the late 1800s. Born in 1909, he was by all accounts a shy young man who liked stamp collecting, taking photographs and keeping pigeons. Although small (he grew to just five foot six), he was a talented fencer. He did not excel academically except in mathematics so his father suggested that, rather than attend university, Nicholas should follow in his footsteps and go into banking.

By 1938 Nicholas was a successful stockbroker and planning a Christmas ski trip to Switzerland. The friend he was travelling with unexpectedly phoned to cancel the Swiss holiday, saying he was in Prague helping refugees who were fleeing Hitler. Winton decided to join him. Terrified by Hitler's advances, thousands of refugees were pouring into the Czech capital. Winton arrived in Prague on New Year's Eve and was appalled at their plight. He spent the first three weeks of 1939 working non-stop. He visited freezing, snow-covered refugee camps where conditions were dreadful. He recalled: 'A lot of refugees were in very bad shape. They felt the days were numbered before the Germans were going to arrive in Czechoslovakia but how could they save themselves? What could they do? Where should they go? They were stuck.'

He made contact with the British Committee for Refugees in Prague. It was doing everything possible to help relocate those it judged to be at greatest risk from the Nazis but Winton quickly realised that there was a special group he might be able to assist: those who had decided to stay in the country but wanted to send their children out of Hitler's reach. No one was helping them. He wrote to his mother, Barbara, in London asking her to go to the British Home Office and find out what guarantees were needed to bring children into Britain. He set up an office at his hotel which was quickly besieged by anxious parents begging him to get their children out of Czechoslovakia. Winton and his small team of helpers began to assemble a list of children's names. His meetings with distraught parents attracted the attention of German security agents stationed in Prague. He noticed he was being followed and took great pleasure in telling the agents to 'piss off' in perfect German. When a beautiful woman struck up a friendship with Winton, his friends warned him that she was a well-known spy. She said her name was Kristen and that she worked for the Swedish Red Cross. Winton must have won her over because his temptress contrived to have thirty children evacuated to Sweden.

It wasn't long before Winton's office in London became impatient about his extended 'holiday' in Czechoslovakia. And when Winton telephoned to request an extra week, his boss asked: 'Why do you want to stay there helping people no one cares about?' Winton stayed the extra week and added more names to his list.

When he returned to London in January 1939, he brought with him hundreds of photos and details of children. He continued to work as a stockbroker but at the end of each business day he rushed home to work on his rescue plan. Late into the night he wrote letters asking the German authorities to allow Czech children to leave and begging other countries to let them in. The only positive response he received was from the British government. Its Home Office had already started a programme to allow unaccompanied Jewish children from Germany and Austria into Britain. It now extended this to include Winton's children from Czechoslovakia. But the British authorities set difficult conditions. He had to find a family to look after each child and £50 had to be deposited to cover the child's eventual return home. It was, said Winton, 'a hell of a lot of money'

(£50 in 1938 is worth around £2,500 today). His was not an official refugee organisation or even an unofficial one with members and officers but he made it look like one. He had notepaper printed with the heading 'British Committee for Refugees from Czechoslovakia – The Children's Section'. Then, said Winton, 'in a very special meeting I had with myself, I appointed myself honorary secretary.'

His next tasks were to find British families to look after the children and money to fund his project. His team in Prague sent more photos of children on their list and Winton made sure they appeared in the British press. Offers of homes and money came flooding in. Winton, ever the businessman, recalled asking those willing to take a child '"whether they wanted a girl or a boy". Then we said "what age?" and we gave them pictures of half a dozen children. Then we asked them to choose a child.' It was, he admitted, a rather commercial approach but 'it was quick, effective and it worked in most cases'. When the Home Office failed to provide the necessary passports and travel documents on time, he set up a printing press and made false ones. Asked about this years later, he replied cheekily: 'We didn't bring in anyone illegally. We just speeded the process up a little.'

Nicholas Winton's first Kindertransport left Prague on 14 March 1939, the day before Germany invaded Czechoslovakia. Vera Diamant remembers waking up on the morning of 15 March and seeing her parents and sister standing by the window, tears flowing down their faces. 'I looked out and there were German soldiers marching into our square. People were standing on the streets silently. Then one person started singing the Czech anthem. Everyone started singing.' A Nazi officer commandeered the best room in Vera's house. He demanded that the family speak German from now on. When her father replied that his family would continue speaking Czech to one another, the officer spat in his face. After that Vera says her mother 'left no stone unturned to find a way to get us out of the country'. And Joe Schlesinger recalls, 'You could feel the danger getting closer and closer and closer.' He heard Hitler's voice on the radio shouting that Jews were no better than rats in the sewers. He was chased through the streets by a group of boys from the Hitler Youth organisation. Hans Feige remembers very clearly the day the Germans invaded and 'German horses, tanks, cars and soldiers marched into our village.' He was terrified when an SS officer in jackboots shouted at his father for cutting down a dead fruit tree. 'He ranted on about Jews saying, "You cause trouble wherever you are."'

The invasion of Czechoslovakia made Winton more determined than ever to bring needy children to Britain. He stepped up his efforts to get publicity for his cause. He launched appeals on the radio, in synagogues, churches, schools. An article in the widely read *Picture Post* magazine brought a huge response from people offering homes to his children. Most host families were Christian which prompted two rabbis to pay Winton a visit. They told him that he must stop sending Jewish children to Christian families. He replied, 'I won't stop and if you prefer a dead Jew in Prague to a live one who is being brought up in a Christian home, that's your problem.'

After the German occupation the team Winton had left behind in Prague were visited by ever-increasing numbers of parents desperate to get their children out of Czechoslovakia. The list grew to 5,000 names. When Vera Diamant's mother found out about the British office organising Kindertransports, she lined up for hours to get her children's names on the list. She didn't tell her husband what she was planning until the day in May that she learned Vera and Eva had been offered places on a train to London. Vera remembers her father's reaction: 'There was a deathly silence. He looked pale and drawn. He buried his face in his hands and said, "All right. Let them go."' Her mother took her and her sister to a dressmaker and had a new wardrobe made. Everything fitted perfectly. Vera says there was no room for growth because: 'The only way my parents could deal with it was to say to themselves and us, "Let's hope it is all a storm in a teacup. You'll be back within a year."'

Joe Schlesinger was thrilled 'to be going on an adventure. Here was I, a kid from central Europe, going to England.' His only worry was that he was going to a place called Newcastle. When he checked a map, he found there were five Newcastles and he wondered if he would end up in the right one. Hans Feige will never forget his father calling him and his brother into the living room and telling them to sit down. He said, 'You are going on a long journey to a country called England. You will have to go alone; we can't come with you.' He then taught his young son four very important words: 'I can't speak English.'

The Kindertransports from Prague to London continued slowly but regularly: one in March, two in April, one in May, one in June, two in July and one in August. The Germans allowed the trains to depart but Winton said it was never easy. 'When transport was due to leave I'd get a message from the Germans: "We can't let the transport go unless you give us so much more money." They were terrible. We just had to find the money.'

Once the trains left Prague, they travelled through Germany to the coast of Holland. Hans Feige recalls 'a huge number of children on board with no one seeming to look after us'. And German soldiers 'scaring us' as they walked up and down the aisles, occasionally stopping to search a child's suitcase then throwing its meagre contents on the ground. Vera Diamant only felt safe when the train crossed the border into Holland. The German soldiers got off and 'Dutch women with beautiful smiling faces gave us chocolate through the windows of the train.' The children crossed the North Sea to England overnight by ship. For Joe Schlesinger, who'd only seen small paddle-steamers on the Danube River, the ferry seemed enormous. Just as he was nodding off to sleep, 'I heard voices from nearby cabins singing the Czech national anthem.' The words 'where is my home?' seemed to him uncomfortably appropriate.

The trains from Prague ended their journeys at London's Liverpool Street station. A reporter described the arrival of the one carrying Joe, Vera and Hans.

> I have seldom seen a more moving sight than the arrival of Czech children last Friday. One of them was less than three years old, the eldest must have been about 15. Policemen kept a gangway for the crocodile which was led off to a gymnasium, curtained down the middle. The children sat on benches on one side of the curtain, the parents were on the other. As each name was called out, the child went through an opening in the curtain and was welcomed by its new parents on the other side.

Not all the foster parents arrived at the station on time. One group of five boys was left sitting on their suitcases for hours. A taxi driver noticed them, took them off for fish and chips and then drove them home to spend the night with his wife and young child in a one-bedroom apartment. Joe Schlesinger's uncle failed to meet the train so he and his brother Ernie were taken off to stay the night with a woman he later learned was Winton's mother. The next day they were put on a train to the right Newcastle. Nine-year-old Vera Diamant was supposed to be going to a family near Liverpool. Her fourteen-year-old sister Eva had been promised a place in a girls' school elsewhere. Eva's name was called and off she went but Vera just went on sitting there 'alone in this big hall'. When her foster mother finally arrived Vera remembers 'a little lady, hardly taller than me who was laughing and crying at the same time. She spoke words I didn't understand but which I later learned were "you shall be loved". And loved I was.' Hans Feige's foster mother was waiting for him but all he can remember is seeing double-decker buses for the first time. 'I was scared stiff. I thought they would topple over.'

Nicholas Winton's largest Kindertransport was supposed to depart Prague on 1 September 1939. That was the day Germany invaded Poland. It was the beginning of the Second World War. All borders slammed shut and the train never left the station. Most of the 250 children on board died in concentration camps. With the start of the war Winton realised his rescue work was over. His mother took over the running of his operation, dealing with any problems the children encountered. Winton turned his talents to Britain's war effort, joining the Royal Air Force and becoming a flight trainer. He seldom talked about the children he saved. When asked why, he said, 'For me it was an episode. So many more important things happened afterwards.' Those 'important things' included working for the International Committee for Refugees after the war. He collected war loot that American troops found in a German concentration camp – everything from fur coats and jewellery to glasses and gold teeth – and supervised its sale to aid Jewish organisations. He worked for the International Bank for Reconstruction and Development arranging loans to war-torn European countries, which was where he met his Danish wife, Grete. They married in 1948. He later went on to raise money for a British mental health charity and for a home for the elderly.

The story of Nicholas Winton's hidden heroism might never have been told without the scrapbook assembled in 1939 by W. H. Loewinsohn, his assistant secretary. After Winton's wife found the scrapbook in 1988 prompting the article in a British newspaper, the popular BBC TV programme *That's Life* picked up the story. It invited Winton – then aged seventy-nine, to appear on the show but didn't tell him that researchers had traced some of the children he had saved. In a magic moment, the woman sitting next to him introduced herself as Vera Diamant. She said she had come to London from Prague on one of his trains in 1939. When the programme's host asked if anyone else in the audience was one of 'Nicky's children' dozens of people stood up. As the normally reserved Winton wiped away tears, he was hugged and kissed by men and women who owed their lives to him.

Nicholas Winton finally got the recognition he deserved. He was dubbed 'Britain's Schindler' and received honours including a knighthood from the Queen and the Czech Republic's highest award, the Order of the White Lion.

Joe Schlesinger became an award-winning news correspondent at the Canadian Broadcasting Corporation. He told us his life changed the day he learned the name of the man who saved him. 'It was breathtaking. Everything fell into place – like finding the missing piece of a jigsaw. Without it there was a hole. It filled the hole, completing the story of my life. This was the man who made it possible for me to survive.'

Vera Diamant, who wrote under her married name Vera Gissing, stayed on in Britain and helped look after Winton in his old age. For her he became 'an honorary father to the biggest family in the world. There are at least 6,000 people round the world who are alive today because of him.'

Hans Feige also made Britain his permanent home. He changed his name to John Fieldsend, converted to Christianity and was ordained an Anglican minister. He refused to return to Czechoslovakia after the war, preferring to look to the future, not back at the past. It was only after meeting Winton that he went back in 1991. He said that learning who had saved his life 'acted as a trigger to find out what happened to my family. I couldn't return to life as usual. I couldn't live with myself without going back.'

Nicholas Winton died in 2015 at the age of 106. He had a motto: 'If something is not impossible, there must be a way of doing it.' It was that thought, he said, that kept him trying to save children's lives when everyone told him he'd never succeed.

Acknowledgements

WE ARE FORTUNATE to have two historians in the family who provided valuable help and ideas for this book. Our son, Dan Snow, author of *Death or Victory: The Battle of Quebec and the Birth of Empire*, came up with inspiring suggestions of potential subjects, as did Ann's sister, Margaret MacMillan, author of *Peacemakers* and *The War that Ended Peace*. Margaret is also an accomplished editor and offered us very welcome literary guidance throughout. We owe thanks too to the hundreds of other friends and acquaintances who have alerted us to people whose stories deserve to be told.

We could not have written this book without the help of our living subjects Major-General Corran Purdon, Mike Sadler, and Ahmad Terkawi, all of whom graciously set aside time to tell us their stories and answer our questions. We are also grateful to those who have provided us with material, pictures and family memories of our earlier subjects. We have detailed their contributions in our *Notes and Sources*.

But we owe a particular debt to Kit Power, descendant of Edward Seager; Margaret Sheil, who's descended from Florentia Sale; Martin King who searched for and found Augusta Chiwy; Sumner Jackson's son Phillip Jackson and granddaughter Loraine Riemer; Eric Newby's daughter Sonia Ashmore; Duncan Dunbar-Nasmith, the submariner's grandson; Jeff Prior, son of Jack Prior, the American doctor who worked with Augusta Chiwy during the Battle of the Bulge; Sir Alistair Horne who met Krystyna Skarbek in Cairo; Joe Schlesinger and John Fieldsend, two of Nicholas Winton's *Kinder*; Captain Tim Lightoller, Charles Lightoller's grandson; Rosemary Vellender, Helen Thomas's granddaughter; Patrick Aylmer, great nephew of Mary Borden; and Hans von Werra, great nephew of Franz von Werra. Thanks are also due to our friends Nancy Durham and Bill Newton Smith for bringing Ruth Werner to our attention, and to the Belgian

Ambassador in London, Guy Trouveroy, for suggesting that we write about Belgium's Second World War hero, Augusta Chiwy.

Finally, our agent, Julian Alexander has been a powerful source of wisdom and advice and we have had the usual highly professional treatment and guidance from our publisher, John Murray. In particular we thank Roland Philipps, who commissioned us to write this book and the rest of the John Murray team: Becky Walsh, Caroline Westmore, Ruby Mitchell, Lyndsey Ng and Nick Davies have done a marvellous job making Roland's vision a reality. For the pictures we have to thank Juliet Brightmore for her painstaking research and her cooperation with Andrew Smith in the design of the cover.

Illustration Credits

ANL/REX/Shutterstock: chapter 12. Photograph © Michael Blyth: 12 above left/Anon, *A Voyage to the South Seas 1745*. Bournemouth News & Pic Service/REX/Shutterstock: chapter 23. Bridgeman Images: 1 above/Florence Nightingale Museum London, 2 below/Granger, 7 and 10 centre left/National Army Museum London, 16 above left and chapter 17/Pictures From History. Cardiff University, Special Collections and Archives: 8 above left and right, chapter 15 right. China News/REX/Shutterstock: 9 below left. Chronicle/Alamy: 5 above left, 11 below left and right. Courtesy of Jane Conway, from her book *A Woman of Two Wars: The Life of Mary Borden*, Munday Books UK, 2010: 15 above. Collection of the Duke of Northumberland: 3 above and chapter 6. Mary Evans Picture Library: chapter 15 left. Paul Fearn/Alamy: 9 centre right. Courtesy of John Fieldsend, from his autobiography *A Wandering Jew*, Radec Press, 2014: 16 above right. John Frost Newspapers/Alamy: 12 centre right. Getty Images: 1 below right, 4 centre right/IWM, 12 below left/ullstein bild, 13 below/PhotoQuest, 16 below/Joseph Eid/AFP, chapters 10, 13, 25, 26/ullstein bild. Granger Historical Picture Archive/Alamy: 2 above, 5 above right and centre right, 13 above, chapter 16. © Imperial War Museum London: 4 above left/HU57120, 15 below/E8308. Courtesy of Martin King: 11 above and chapter 20. *A Week at Waterloo in 1815: Lady de Lancey's Narrative*, John Murray, 1906: 10 above right, chapter 18. Library of Congress Washington DC: chapter 2. Paul Martin News/Alamy: 12 below right. Mirrorpix/Alamy: chapter 7. Courtesy of Martin Nasmith's family: chapter 22. © National Portrait Gallery London: chapter 29. Estate of Eric Newby: 14 below left and right and chapter 27. Niday Picture Library/Alamy: 5 below, chapter 4. PA Images: chapters 3/Fiona Hanson, 31. Copyright Pitt Rivers Museum, University of Oxford: 6 below/PRM 1998.309.1940. Private

Collections: 6 above left and right, 9 above left, 10 below, chapters 1, 5, 11, 19, 30. Courtesy of Major-General Corran Purdon: chapter 9. © The Royal Aeronautical Society (National Aerospace Library)/ Mary Evans Picture Library: 14 above. © Royal Academy of Arts London/photo Prudence Cuming Associates: 3 below/detail from cartoon, *The Meeting of Wellington and Blücher After the Battle of Waterloo* by Daniel Maclise, 1858–9. Royal Green Jackets (Rifles) Museum: chapter 14 left. Courtesy of Mike Sadler: chapter 8. *The Sun*/News Syndication: 1 below left. *The Autobiography of Lieutenant-General Sir Harry Smith*, John Murray, 1902: chapter 14 right. Peter Snow: chapter 28. Alexis Soyer: 8 below/*Soyer's Culinary Campaign*, G. Routledge & Co., 1857. TopFoto: 4 below. WorldPhotos/Alamy: chapter 24. Yale Center for British Art, Paul Mellon Collection: chapter 21/detail from a painting by Charles Brooking *c.*1750.

Notes and Sources

Chapter 1: Edward Seager

Edward Seager's letters were lent to us by his descendant Kit Power in April 2015. We had a lot of further informative material from another descendant, Michael Seager of Adelaide, in Australia. Nigel Harvey of the museum of the 8th King's Royal Irish Hussars in Eastbourne, Sussex, added some valuable research. The regiment's history is told in the *History of the 8th King's Royal Irish Hussars 1693–1927*, vol. 2, by Robert H. Murray (W. Heffer, Cambridge, 1928). A good history of the Crimean War is *Crimea: The Last Crusade* by Orlando Figes (Penguin, London, 2010). There are some fascinating letters from other Light Brigade men in Anthony Dawson's *Letters from the Light Brigade* (Pen and Sword, Barnsley, 2014). Mark Adkin has written a detailed account of the famous charge in *The Charge* (Pimlico, London, 2004). Fanny Duberly's *Crimean Journal* (classictravelbooks.com, USA, 2006) provides a very conversational commentary on the Light Brigade.

Chapter 2: Joshua Chamberlain

There are a number of books about Chamberlain. The most colourful – though not necessarily the most accurate – is his own *The Passing of the Armies* (Bantam, New York, 1992). A good account of Chamberlain's life is John Pullen's *Joshua Chamberlain: A Hero's Life and Legacy* (Stackpole Books, Mechanicsburg, Pennsylvania, 1999). A more critical account of Chamberlain's career is by Edward G. Longacre, *Joshua Chamberlain: The Soldier and the Man* (Combined Publishing, Conshohocken, Pennsylvania, 1999). For detailed maps, *The Civil War Battlefield Guide*, edited by Frances H. Kennedy (Houghton Mifflin, Boston, 1990) is very well stocked.

Chapter 3: Private Johnson Beharry VC

Beharry wrote his own very racy biography with Nick Cook: *Barefoot Soldier* (Sphere Books, London, 2006). The *Daily Telegraph* of 17 May 2010 carried the news of Beharry's attempted suicide.

Chapter 4: Paul Revere

By far the most comprehensive book about this American hero is *Paul Revere's Ride* by David Hackett Fischer (Oxford University Press, 1994). An excellent account of the times during which Revere lived can be found in *Revolutionary America 1763–1789* by Ronald M. Gephart (Library of Congress, Washington, DC, 1984). Paul Revere's own account of his ride is contained in depositions in the *Revere Family Papers* in the library of the Massachusetts Historical Society. The poem that immortalised Revere written by Henry Wadsworth Longfellow was first published in the January 1861 issue of the *Atlantic Monthly* under the title 'Paul Revere's Ride'. It was later re-named 'The Landlord's Tale'.

Chapter 5: Laura Secord

Of the many books written about Secord, the two that appealed to us were: *The Story of Laura Secord and Canadian Reminiscences* by Emma A. Currie (William Briggs, Toronto, 1900); and Ruth McKenzie's *Laura Secord: The Legend and the Lady* (McClelland and Stewart, Toronto, 1971). Secord's own accounts of her deed can be found in two letters she wrote to Gilbert Auchinleck, a historian of the War of 1812, quoted in articles for the *Anglo-American Magazine* (Toronto, 1853).

Chapter 6: Harry Percy

Brian Cathcart's excellent book *The News from Waterloo: The Race to Tell Britain of Wellington's Victory* (Faber & Faber, London, 2015) sheds important new light on Harry Percy's trip from Waterloo to London. So does *Waterloo: The Aftermath*, by Paul O'Keeffe (Bodley Head, London, 2014). We also consulted the Rothschild archives in the City of London. Julian Charles

Young spoke with Mrs Boehm sixteen years after her ball and recorded her remarks in his journal, cited in the *Spectator*, Letters to the Editor, 20 January 1894.

Chapter 7: Krystyna Skarbek

We first read of Skarbek's intriguing adventures in *The Heroines of SOE* by Squadron Leader Beryl E. Escott (History Press, Stroud, 2010). Two other books, *Christine: SOE Agent and Churchill's Favourite Spy* by Madeleine Masson (Virago, London, 1995), and *The Spy Who Loved* by Clare Mulley (Macmillan, London, 2012), made her an irresistible subject. We also learned about Krystyna first-hand when invited to lunch with historian Sir Alistair Horne in April 2015. He told us about meeting her in Cairo when he worked for MI6 there after the First World War.

Chapter 8: Mike Sadler

We interviewed Mike Sadler in his Cheltenham apartment on 12 May 2015. Ben Macintyre writes much about Sadler in his excellent history of the Special Air Service, *SAS: Rogue Heroes* (Viking, London, 2016). There's useful material about Sadler's role in the *SAS War Diary 1941–1945* (Extraordinary Editions, London, 2011). For the extraordinary story of Paddy Blair Mayne see the detailed account in *Rogue Warrior of the SAS* by Martin Dillon and Roy Bradford (Mainstream, Edinburgh, 2007).

Chapter 9: Corran Purdon

We interviewed Major-General Corran Purdon at his West Country home in April 2015. Purdon's own story is also retold by Sean Rayment in *The Suicide Raid* (HarperCollins, London, 2013). The story of the St Nazaire raid is recounted in *St Nazaire Raid: Operation Chariot* by James Dorrian (Pen and Sword, Barnsley, 2006).

Chapter 10: Benedict Arnold

The best account of the whole conspiracy is in Nathaniel Philbrick, *Valiant Ambition: George Washington, Benedict Arnold and the Fate of the American Revolution* (Viking, New York, 2016). Another competent account is James Thomas Flexner's *The Traitor and the Spy* (Syracuse University Press, New York, 1991).

Chapter 11: Sonya

This unusual spy told her own story in *Sonya's Report* using the pen name Ruth Werner (Chatto and Windus, London, 1991). Her son Michael Hamburger added his personal recollections in a BBC Radio documentary *Codename Sonya* broadcast in 2003. The whole story is laid out as a tale of frightening conspiracy in *Treachery: Betrayals, Blunders and Cover-Ups: Six Decades of Espionage* by Chapman Pincher (Mainstream, Edinburgh, 2011). Max Hastings in his *The Secret War* (William Collins, London, 2015) sets Sonya's career in context.

Chapter 12: Ursula Graham Bower

Bower wrote a memoir called *Naga Path* (John Murray, London, 1950), describing her life as a British army guerrilla in north-east India. She also did a lengthy and revealing television interview with anthropologist and historian Professor Alan Macfarlane in 1985 which can be seen on YouTube. *The Naga Queen: Ursula Graham Bower and her Jungle Warriors 1939–45* by Vicky Thomas (History Press, Stroud, 2012) uses letters and diaries supplied by Bower's daughters to enrich the story still further. Fergal Keane's *Road of Bones: The Epic Siege of Kohima 1944* (Harper Press, London, 2010) provides important details of the fight in which Bower found herself immersed.

Chapter 13: Sir Robert and Lady Sale

Lady Sale's diary was published by John Murray in 1843. See also *Retreat and Retribution in Afghanistan, 1842* by Margaret Kekewich (Pen and Sword,

Barnsley, 2011), and Lady Sale, *The First Afghan War (Military Memoirs)*, edited by Patrick MacRory (Longmans, London, 1969). Margaret Sheil, Lady Sale's great-great-great-granddaughter gave us access to some useful family letters and other documents in her possession. The story is set in context superbly by William Dalrymple in *Return of a King: The Battle for Afghanistan* (Bloomsbury, London, 2013).

Chapter 14: Harry and Juana Smith

Harry Smith tells his own story in *The Autobiography of Lieutenant-General Sir Harry Smith*, edited by G. C. Moore Smith (John Murray, London, 1902). John Kincaid, a rifleman and friend of Harry and Juana Smith, writes about them in *Tales from the Rifle Brigade* (Pen and Sword, Barnsley, 2005). Harry Smith also figures prominently in *To War with Wellington* by Peter Snow (John Murray, London, 2010) and *When Britain Burned the White House* by Peter Snow (John Murray, London, 2013). Mark Urban's *Rifles* is a superb narrative of the experiences of Harry Smith's regiment (Faber & Faber, London, 2003).

Chapter 15: Helen and Edward Thomas

Helen Thomas's unforgettably moving account of life with her beloved husband Edward was published in two books by Heinemann. The first, *As It Was*, came out in 1926 and the second, *World Without End*, in 1931. Helen Thomas's granddaughter, Rosemary Vellender, shared her memories with us.

Chapter 16: Alexis Soyer

Relish: The Extraordinary Life of Alexis Soyer by Ruth Cowen (Weidenfeld and Nicolson, London, 2006) whetted our appetite for the story of this eighteenth-century celebrity chef. Context was added by William Howard Russell's ground-breaking reports for *The Times* assembled in *Despatches from the Crimea* (Frontline Books, Barnsley, 2008). We also found references to Soyer in the letters Edward Seager sent home to his family – see Seager notes above.

Chapter 17: Dr Norman Bethune

There are several comprehensive biographies of Bethune including Ted Allan's and Sydney Gordon's *The Scalpel, the Sword* (McClelland and Stewart, Toronto, 1952), Roderick Stewart's *Bethune* (New Press, New York, 1973) and *Norman Bethune* by Adrienne Clarkson (Penguin Canada, Toronto, 2009). Bethune's beleaguered colleague Jean Ewen also recorded her memories of the great man in *China Nurse 1932–1939* (McClelland and Stewart, Toronto, 1981). Others who worked with or knew Bethune spoke about him in the National Film Board of Canada documentary directed by Donald Brittain in 1964.

Chapter 18: Magdalene de Lancey

Magdalene de Lancey tells her story in *A Week at Waterloo in 1815* (John Murray, London, 1906). She records her wedding date as March 1815.

Chapter 19: Sumner Jackson

We interviewed Sumner Jackson's son, Phillip, in Paris in 2015. Jackson's story is well told by Charles Glass, *Americans in Paris: Life and Death Under Nazi Occupation 1940–1944* (HarperCollins, London, 2009). Also Hal Vaughan in *Doctor to the Resistance* (Brassey's, Washington, DC, 2004). Alex Kershaw described Nazi-occupied Paris in *Avenue of Spies* (Random House, New York, 2015). There was a useful article in *World War II* magazine by Stephen Budiansky dated 30 July 2010.

Chapter 20: Augusta Chiwy

The Belgian ambassador to London, Guy Trouveroy, first alerted us to Chiwy. He put us in touch with British historian Martin King whose determined efforts to track down the elusive nurse are chronicled in his award-winning television documentary *Searching for Augusta: The Forgotten Angel of Bastogne*. King first learned about Chiwy while researching a book, *Voices of the Bulge*, and, by finding her, has assured her place in military history. We contacted Dr Jack Prior's son Jeff who provided details about his father's work with Chiwy. The Battle for Bastogne is set in context by Antony Beevor in *Ardennes 1944: Hitler's Last Gamble* (Viking, London, 2015), along with *Band of Brothers* by Stephen E. Ambrose (Simon and Schuster, London, 1992).

Chapter 21: John Bulkeley

John Bulkeley's joint narrative with John Byron's is retold from the first editions in 1743 and 1768 and edited by Alan Gurney in *The Loss of the Wager* (Boydell Press, Woodbridge, Suffolk, 2004).

Chapter 22: Martin Nasmith

Nasmith's grandson, Duncan Dunbar-Nasmith was very helpful, sharing with us research he has done into his grandfather's career. We have also consulted: Peter Shankland and Anthony Hunter, *Dardanelles Patrol* (Collins, London, 1964); C. G. Brodie, *Forlorn Hope 1915* (Frederick Books, London, 1956); and William Carr, *By Guess and by God* (Hutchinson, London, 1930).

Chapter 23: Charles Lightoller

Lightoller tells his own story in detail in *Titanic and Other Ships* (Oxford City Press, Oxford, 2010). His life is well documented too in Patrick Stenson, *Lights: The Odyssey of C. H. Lightoller* (Bodley Head, London, 1984). We also spoke to Captain Tim Lightoller RN, who remembered navigating *Sundowner* with his grandfather when he was a young boy.

Chapter 24: Billy Bishop

Billy Bishop's own biography *Winged Warfare* is edited by Stanley Ulanoff (Bailey Bros and Swinfen, London, 1978). There is a useful account by Dan McCaffery, *Billy Bishop: Canadian Hero* (James Lorimer, Toronto, 2002). There is also a good account of Britain's Royal Flying Corps and early Royal Air Force in *The Royal Flying Corps in World War I* by Ralph Barker (Constable, London, 1995).

Chapter 25: Butch O'Hare

The life of O'Hare is explored in satisfying detail in *Fateful Rendezvous* by Steve Ewing and John B. Lundstrom (Bluejacket Books, Annapolis, 2004). For context we turned to *Carrier War* by Lieutenant Oliver Jensen (Simon and Shuster, London, 1945), *History of United States Naval Operations in World War II* by Rear Admiral Samuel E. Morison (Little Brown, London, 1948) and *The Man Who Got Capone* by Frank Spiering (Bobbs-Merrill, Indianapolis, 1976).

Chapter 26: Franz von Werra

We talked to Franz von Werra's great-nephew, Hans von Werra, eighty-three years old in 2016. Hans told us of his memories of his great-uncle and sent us some pictures of him as a young Luftwaffe pilot. We found a useful biography of von Werra: *The One That Got Away* by Kendal Burt and James Leasor (Collins with Michael Joseph, London, 1956), adapted for the film of the same title starring Hardy Krüger. There is an article in the *Ogdensburg Journal* (27 January 1941) describing how von Werra was feted in New York after he'd arrived in the USA.

Chapter 27: Eric Newby

Newby tells his own story with a wry sense of humour in *Love and War in the Apennines* (William Collins, London, 2010). We spoke to Newby's daughter, Sonia, who was most helpful with recollections about her father and showed us photos and other related material.

Chapter 28: Ahmad Terkawi

We interviewed Ahmad Terkawi and his wife in Sweden in October 2016. We agreed not to name his wife or children in our account in case they want to return to Syria.

Chapter 29: Mary Borden

Borden wrote extensively about her wartime experiences in her poetry and in her recollections about nursing in the First World War in *The Forbidden Zone* (William Heinemann, London, 1929). Borden's *Journey Down a Blind Alley* (Hutchinson, London, 1946), describes her work in the Second World War. Jane Conway has written a first-rate biography of Borden called *A Woman of Two Wars: The Life of Mary Borden* (Munday Books, UK, 2010). It is also worth reading works by Borden's second husband, Sir Edward Spears. They include *Liaison 1914* (Eyre & Spottiswood, London, 1930) and *Prelude to Victory* (Jonathan Cape, London, 1939). *To War with Whitaker: The Wartime Diaries of the Countess of Ranfurly 1939–45* by Hermione, Countess of Ranfurly (Heinemann, London, 1995) provides useful context.

Chapter 30: John Rabe

The Good German of Nanking: The Diaries of John Rabe (Little, Brown, London, 1999) provided us with a vivid day-by-day account of Rabe's activities. *The Rape of Nanking: The Forgotten Holocaust of World War II* by Iris Chang (Penguin Books, London, 1998) sets the grim context.

Chapter 31: Nicholas Winton

We interviewed two of 'Nicky's children', Joe Schlesinger and John Fieldsend. Vera Gissing did a lengthy interview for the Shoah Foundation in 2012 which can be found on YouTube. Gissing was a great diary writer and recollections of her childhood can be found in her book *Pearls of Childhood* (Pavilion Books, London, 2007). John Fieldsend described his experiences in *A Wandering Jew* (Radec Press, Thame, Oxon, 2014). Joe Schlesinger's autobiography is entitled *Time Zones: A Journalist in the World* (Random House, Toronto, 1990). Barbara Winton has written a book about her father called *If It's Not Impossible . . . The Life of Sir Nicholas Winton* (Matador, Leicester, 2014). The BBC's 1998 *That's Life* programme that revealed Winton's compassion to the world can be seen on YouTube. Slovak filmmaker Matej Minac has made three films about Winton's work: a drama, *All My Loved Ones* (1999) with Rupert Graves playing Winton; the documentary *The Power of Good: Nicholas Winton* (2002), which won an Emmy Award; and the documentary drama *Nicky's Family* (2011).

Index